Robert E. Eliason

Early American Brass Makers

Brass Research Series: no. 10
Stephen L. Glover, Editor

The Brass Press
a division of
Editions Bim
P.O. BOX 300 - CH-1674 VUARMARENS / SWITZERLAND
✆ ++41 (0)21 909 10 00 - Fax ++41 (0)21 909 10 09
email: order@editions-bim.com - web: http://www.editions-bim.com

Library of Congress Cataloging in Publication Data

Eliason, Robert E
Early American brass makers.

(Brass research series ; no. 10 ISSN 0363-454X)
1. Brass instruments--Makers--United States--
Biography. I. Title. II. Series.
ML404.E47 788'.01'20922 [B] 79-11880
ISBN 0-914282-25-5 pbk.

Second (revised) printing, 1981.

Editions Bim The Brass Press
P.O. BOX 300, - CH-1674 VUARMARENS / SWITZERLAND, ✆ ++41 (0)21 909 10 00 - Fax ++41 (0)21 909 10 09
email: order@editions-bim.com - web: http://www.editions-bim.com

Ref. BrP 9

Foreword

Unrecognized by most historians and almost unknown to Americans, the United States had a thriving musical instrument industry in the middle years of the 19th century. Brass bands became the rage throughout the country soon after 1840 and dozens of American makers of brass instruments set about supplying the demand for these products. Among the many makers, five who stand out for the importance of their contributions to the industry are Samuel Graves, Thomas D. Paine, J. Lathrop Allen, Elbridge G. Wright, and Isaac Fiske. Of these, Samuel Graves, whose manufactory was the earliest and most important in terms of production, size and variety of instruments, has been treated in a separate booklet: *Graves & Company Musical Instrument Makers* (Dearborn, Michigan: The Edison Institute, 1975); and a subsequent article: "Recently Discovered Information About Graves & Company, Musical Instrument Makers," *The Herald*, V (April, 1976) pp. 58-63. Thomas D. Paine has also been treated in an article in *The Herald*, V (October, 1976) pp. 15-25, which is presented here with recent revisions. Other chapters on the remaining three makers—Allen, Wright, and Fiske—complete this survey of important 19th century American brass instrument makers.

Each of the four makers discussed contributed some significant invention or improvement toward the development of brass instruments. Thomas D. Paine was the first to produce rotary valves with string linkage—a combination which dominated American production for nearly half a century and is still in use on modern French horns. J. Lathrop Allen was distinguished for his invention of the Allen valve, popular among American players until late in the century. E.G. Wright carried the development of the key bugle to its zenith and was famous for the beautiful presentation instruments he made of silver and gold. Isaac Fiske was noted for his work on eliminating restriction in the bore of brasses and for attempts to provide the cornet with lighter, quicker valve action.

Aside from the technical improvements many of the instruments of this period are beautiful in form and color, and show the excellent handcraftsmanship of early American shop industries.

Thomas D. Paine

Of the many inventions tried and patents obtained by American brass instrument makers before 1875, only one improvement found immediate and lasting success. This was the rather simple innovation of turning a rotary valve with something similar to a bowstring. The advantages of string linkage were quieter operation, simplicity, and better leverage. The concept is still in use on many modern instruments, particularly French horns.

Although music historians have always credited American makers with this invention, it was not known until recently who first utilized the method. The earliest patent mentioning string linkage is United States patent 12,628 of April 3, 1855, granted to Gustavus Hammer of Cincinnati, Ohio. However, in a book published in 1853[1], the Dodworths, a family of well-known bandmasters in New York, claimed to have originated the idea.

The earliest instruments equipped with this improvement discovered to date were not made by either of these inventors, but by Thomas D. Paine & Co. of Woonsocket, Rhode Island. Furthermore, Paine's instruments show both the fully developed application of the idea and an earlier less refined version.

Paine was also the inventor of a unique rotary valve for brass instruments, patented in 1848. His inventions and improvements place him among the five foremost makers of brass instruments in mid-nineteenth century America. During his long life Paine was also a musician and composer, a jeweler, and a violin maker.

Thomas Dudley Paine was born in Foster, Rhode Island, October 9, 1812, the fifth of ten children in the family of John and Polly Paine. John Paine was a schoolteacher, justice of the peace, and sometimes a mechanic. He was evidently well liked and capable for he later served as representative in the Smithfield General Assembly.[2] He may also have been a musician for it is said that Thomas inherited his musical abilities from his father. At the age of ten Thomas was sent to live with an aunt, Surea Bates, in Woonsocket where he began his working career in the spinning room of the Lyman cotton mill. For the next ten years he was employed at the mill, eventually becoming overseer of the spinning room operations. Although the work was not hard, it was tedious, and long hours were the rule. It was said that "fighting and huckleberrying" were the chief sources of "amusement"[3] in Woonsocket, but there must have been music as well, for somewhere in these years Thomas Paine learned to play the violin.

About 1832 John Paine moved to Woonsocket and acquired a small tract of land on the south side of the road from Woonsocket Falls to Union Village.[4] The property already had a dwelling on it and within a year John Paine added a small shop which city records indicate he rented to a watchmaker and a grocer.[5]

In this same year Thomas quit his job at the Lyman mill and apprenticed to Daniel Hubbard, a watch and clock repairman in Woonsocket. In 1834 he married the first of three wives, Perly C. Thayer.

Throughout these years Thomas was also playing his violin for dances. The following appeared in the *Woonsocket Patriot* from October 4, 1834, to February 7, 1835.

> **Cotillion Parties**
>
> The subscribers would respectfully inform the young ladies and gentlemen of this village and vicinity that they have engaged the hall of Col. Holbrook on Tuesday evening October 14 and will continue to attend every other Tuesday evening through the fall and winter for the accommodation of cotillion parties. Hours of attendance from 6 to 12 o'clock. Tickets to be had at the bar of the hotel – price 75¢.
>
> Thomas D. Paine
> Hiram J. Eddy

The end of Paine's apprenticeship in the watch and clock business is marked by an ad appearing in the *Woonsocket Patriot* from April 4 through June 6, 1835:

> **New Establishment**
>
> The subscriber takes this method to inform the inhabitants of Woonsocket Falls and vicinity that he has taken the shop formerly occupied by Mr. Daniel Hubbard, 2nd door north of Whitcomb's Hotel, where he is prepared to clean and repair watches and clocks, and jewelry – and will warrant his work to be done as well and as cheap as at any other shop in this region.
>
> Thomas D. Paine

On March 31, 1837, another *Patriot* advertisement indicated that Hubbard had returned but made no mention of Paine.

Paine's interest in brass making was probably kindled by the enormous popularity of the first brass bands formed in the years between 1835 and 1840. The American Brass Band of nearby Providence presented their first concert January 26, 1838, and the interest aroused could have persuaded Paine that there would be a growing market for brass instruments. With the return of Daniel Hubbard, Paine's watch and clock repair business either ceased or was considerably reduced, stimulating another attempt at something new.

Paine went to Boston for a year in 1840 or 1841. The following listing appeared in the Boston City Directory of 1841: "Thomas D. Paine musical instrument maker home 40 Marion." He seems to have worked at least some of the time with E.G. Wright, the foremost brass maker in Boston at that time. Both Paine and Wright entered key trumpets in the third Massachusetts Charitable Mechanic Association exhibition held at Boston the last of September 1841. The judges reported only that "these instruments came in too late for examination."

Paine returned to Woonsocket in 1842, for an announcement appeared in the *Woonsocket Sentinal & Tomsonian Advocate* on October 12, 1842:

> **Cotillion Party**
>
> The ladies and gentlemen of Woonsocket and vicinity are respectfully invited to the assembly room of Mr. R. Smith on the evening of the clam bake (tomorrow) the 13th instant.
>
> Music – T. D. Paine, 1st violin; E. A. Paine, 2nd do and posthorn; A. Coe, clarionet; G. W. Gates, trombone and ophicleide.
>
> 'Let those dance now who never danced before, and those who've always danced now dance more.'
>
> Tickets 50 cents. Dancing from 6 until 2 o'clock.
>
> N.B. Those in want of music for balls or parties will please address T. D. Paine, Woonsocket.

Paine's obituary in the *Woonsocket Evening Call*, June 3, 1895, mentions that he composed a considerable amount of dance music. Although much of it was probably for his own use as a dance teacher and musician, at least one piece is known to have been published. *The People's Quadrille* by T. D. Paine (figure 1) appeared in *The Instrumental Musician No. 2*, a collection "containing a large number of Marches, Quicksteps, Waltzes, Hornpipes, Contradances, Cotillions, etc."

Probably in 1842 and at least by 1844 Thomas D., his father John, and his brother Emery A., formed Thomas D. Paine & Co. to make brass instruments. The 1844 Providence Almanac lists them as "Music sellers and music instrument makers." They worked in the small shop that John Paine had built next to his house in 1832 or 1833.

An advertisement in the March 12, 1845, issue

Fig. 1 – The People's Quadrille by T. D. Paine published in *The Instrumental Musician* at Boston by Elias Howe, Jr., in 1843. (Photo courtesy of the Music Division, Library of Congress)

of the *Providence Gazette* stated that the new Prodence Brigade Band would be ready for engagements on the first of May. "The wind instruments, 12 in number, are to be upon the improved plan, operating with valves." Another advertisement for the Brigade Band stated that "the cornetts, bassonetts, tubas &c. were all manufactured by T.D. Paine & Co., Woonsocket, which for beauty and sweetness of tone, are not equalled by any in the United States."[6]

"Bassonett" seems to be a term Paine used to describe his baritone, tenor, and alto instruments. Another ad for the Providence Brigade Band in the *Providence Gazette*, May 22, 1845, gives the following description of the bassonett:

> The instruments of this band comprise the valve tubas, and Bassonetts, being a newly invented instrument, which far exceeds the bugles, trombones and ophicleides for richness of tone and convenience of execution.

In 1848 Paine devised and patented the unique rotary valve that brought him attention from many well known musicians. Paine's idea was to reduce the movement and force necessary to operate each valve so that it could be manipulated easier and faster. To this end he constructed rotary valves with an additional passage through the center of

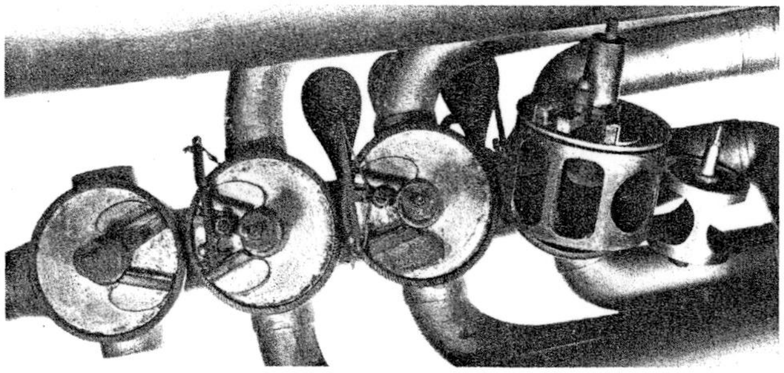

Fig. 2 – Detail of figure 6, tuba in F by Thomas D. Paine, showing Paine's rotary valves. A comparable Vienna rotary valve rotor is shown beside the Paine rotor. (Photo courtesy of the Henry Ford Museum, Dearborn, Michigan)

the rotor (figures 2 and 3). Although the rotor was larger in order to accommodate the extra passage, it needed to move only 1/8 turn instead of the usual 1/4 turn. The patent, No. 5919, November 1848, states that "the advantage of this third or center pipe consists in the shortness, quickness and ease of the action in throwing the wind and sound from the open instrument into the valve crook . . ."

Very soon after the patent Paine redesigned his valve by bending the two outside passages of the rotor 90 degrees toward one end. This made possible a smaller, more compact rotor (figures 7, 8, 9, and 11). All but two of the surviving Paine instruments have valves with this improvement.

The patent text mentions nothing about string linkage nor is any hint of this idea shown in the drawings (figure 3), yet, curiously, every one of Paine's instruments known has some version of it. Even the earliest, dating from before 1848, with valves almost exactly like the patent, has string linkage (figure 2).

A possible explanation is that the Dodworths initially suggested this idea as claimed in the book, *Dodworth's Brass Band School*. Paine, then, was the first to make practical use of the method but could not include it in his patent.

The Dodworths were influential among musicians and were very helpful in spreading the news of Paine's work. The following is from an article, "Brass Bands," by Allen Dodworth which appeared in the magazine *Message Bird*, June 15, 1850, page 361.

> Many are under the impression that a brass instrument that is not imported cannot be good; this is very erroneous, as it is a question whether as good instruments are made in the world as can be, and are made in this country . . .
>
> . . . who would think that a better valve instrument can be made in the little village of Woonsocket, Rhode Island, than in Paris or London! But such is unquestionably the fact. There is a firm of two or three brothers there (T. D. Paine & Co.), who are self taught, in a measure, and who have lately made a number of instruments, with improved valves and other additions which are pronounced by all to be superior to everything yet made, in shape of brass instruments, with regard to workmanship, quality of tone and correctness of tune . . .

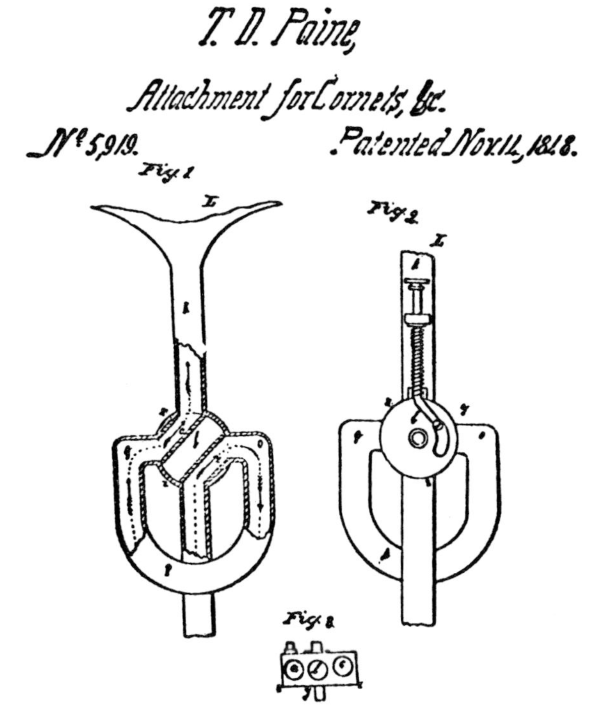

Fig. 3 – Drawing from United States patent 5919 granted to T. D. Paine, November 14, 1848.

Allen Dodworth makes no mention of string linkage here although it would seem natural for him to do so if Paine was making something he, his father, or his brothers had suggested. Perhaps, like so many things, the idea seemed trivial at the time and only became important later when the full range of its possibilities was recognized.

In any case, Paine deserves a large share of the credit for this invention. He was not only the first to apply the idea, but developed the improvement that made it applicable to all rotary valves. In the earlier version (figure 2), the string connects the valve lever to a lever turning the valve rotor. The flexibility of the string reconciles the slightly different arcing motions of the two levers it connects.

This was adequate for Paine's valves which rotated only 1/8 turn, but would not have worked

Fig. 4 – Detail of figure 11, tuba in C by Thomas D. Paine, showing Paine's rotary valves. On the right are three valves using the earlier version of string linkage: strings attached directly to levers turning the valves. At the bottom are three valves showing the improved model with strings wrapping once around the collar on each rotor shaft. (Photo courtesy of the Rhode Island Historical Society, Providence)

well on valves moving 1/4 turn as all other rotary valves do. In the improved version the string wraps once around the collar attached to the rotor shaft, changing the lateral motion of the key lever efficiently to circular motion. It is simple, quiet, and provides maximum leverage at any position, no matter how far the rotor must turn (figure 4). It is this refinement of the idea that rapidly became standard on all American-made rotary valves, and which is still in use today.

The publicity about these new inventions brought the firm of Thomas D. Paine & Co. orders for instruments from several bands. Both the Dodworth Band of New York and the American Brass Band of Providence are known to have used some Paine instruments. Although a violinist primarily, Thomas D. Paine began playing one of his brass instruments in the American Brass Band. A program of March 10, 1851, lists him as the last of three bass, or tuba, players.[7]

After the brief appearance of a key trumpet at

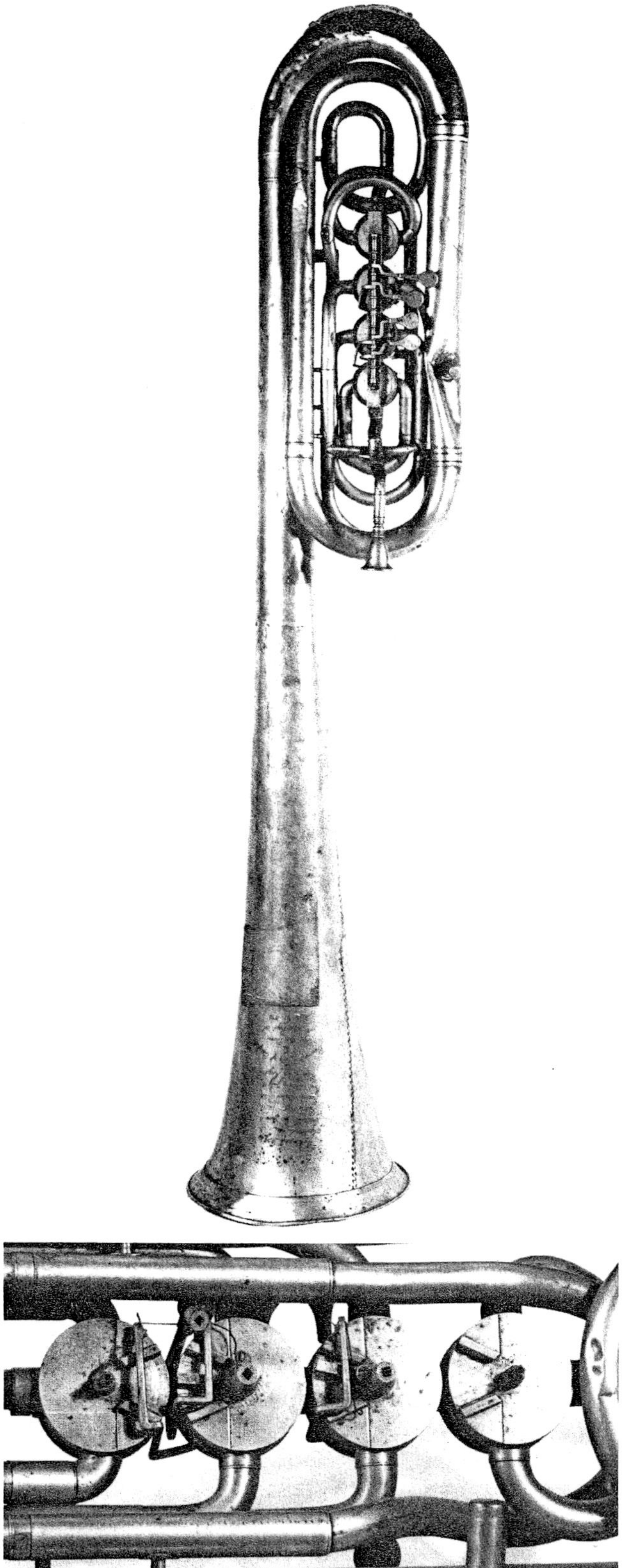

Fig. 5A, B – Overshoulder tuba in E-flat by Graves & Co., Boston, 1851-1854. German silver. Four early-type Paine rotary valves. (Photos courtesy of the Berkshire Museum, Pittsfield, Massachusetts)

the Massachusetts Charitable Mechanic Association exhibit in 1841, Paine's instruments are not known to have been entered in any mechanic exhibitions until 1852. In that year, the following appeared in the *Report of the 22nd Exhibition of American Manufacturers Held in the City of Philadelphia* by the Franklin Institute, page 22:

> No. 2751 A full set of 12 pieces German silver cornet instruments, rotary valves with extra crooks and keys, by Thomas D. Paine & Co. of Woonsocket, Rhode Island, deposited by Beck's Philadelphia Brass Band. These instruments possess the following necessary qualifications in a very eminent degree, viz: superiority of finish, purity and correctness of tone and the important improvement in the valve affords greater facility for execution than any now in use; they well deserve the award of a 1st Premium.

Evidence has recently been found that at least one other maker tried producing Paine valves or perhaps bought valves to use on his own instruments. There is an overshoulder tuba in the Berkshire Museum, Pittsfield, Massachusetts, signed "Graves & Co., 18 Harvard Pl., Boston", with four early-type Paine valves (figure 5). Graves & Co. were at that address from 1851 to 1854 and were well known for their work with various types of valves.

So far seven of Thomas Paine's brass instruments as well as one by his brother Emery, have been found. Since each tells something about the maker and his work they will be examined individually, approximately according to date.

What appears to be the earliest of the surviving instruments by Paine is a tuba in F (air column 380 cm) found in the D.S. Pillsbury Collection at the Henry Ford Museum (figure 6). It is surprisingly large in bore and is not based on proportions of the saxhorns or any known European tubas of that time. Its bell and bore proportions correspond fairly accurately, however, to that of the common American key bugle, and it is a safe guess that Paine took his dimensions from that source. In fact all of Paine's instruments, except the C cornet (figure 9), show this relationship to key bugle proportions.

The construction of the Ford Museum tuba indicates a certain lack of skill not noticed on other instruments by Paine. The bell displays some crinkling which seems to have occurred when it was shaped, and the joint lengthways on each section of tubing is irregular and large. The finish has not been polished to a high gloss and remains dull. There are four Paine valves on this instrument very much like the patent of 1848 except that the

Fig. 6 – Tuba in F by Thomas D. Paine, Woonsocket, Rhode Island, circa 1847. German silver. Four early-type Paine valves. Figure 2 shows these valves in close-up. (Photo courtesy of the Henry Ford Museum, Dearborn, Michigan)

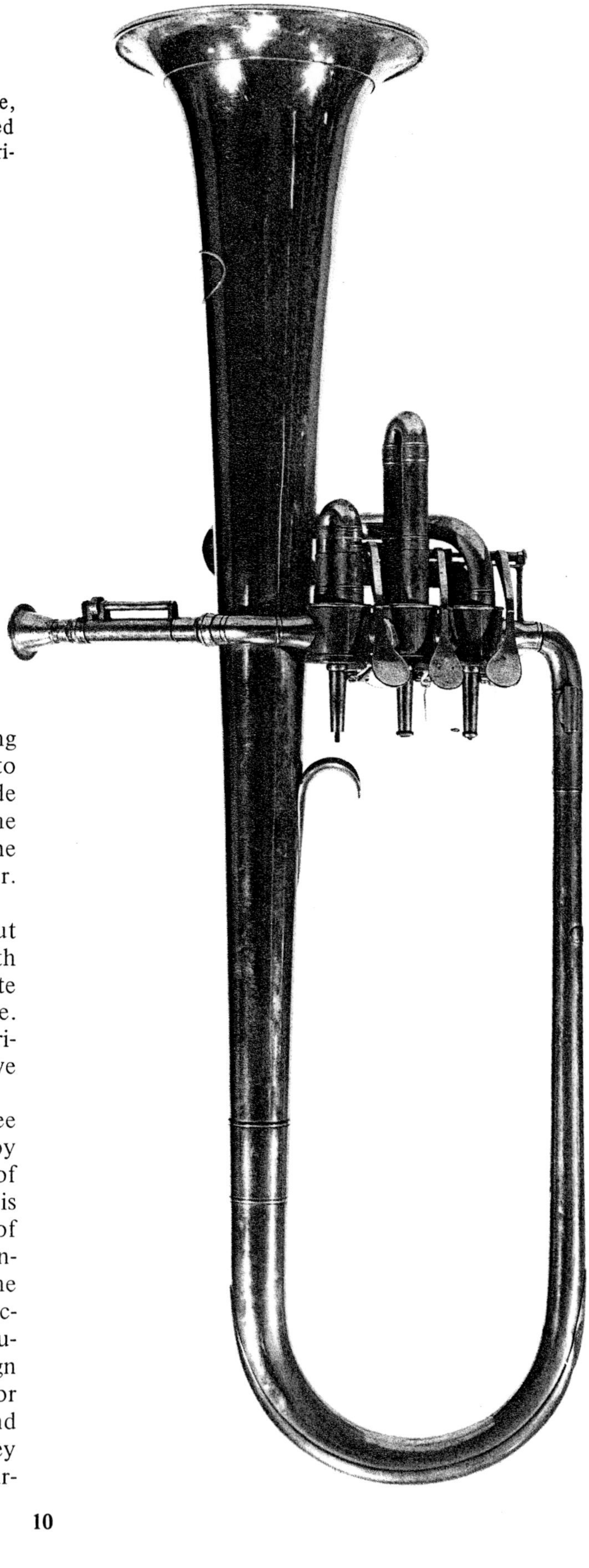

Fig. 7A, B – Alto horn in A-flat by Thomas D. Paine, Woonsocket, 1848. German silver. Three Paine improved model valves. (Photos courtesy of the Rhode Island Historical Society, Providence)

rotors are turned by levers connected by string linkage. Although the shortest valve appears to be first, the finger levers are arranged to provide the usual order of fingering. Unfortunately, the fourth valve tube is missing. Like most other Paine instruments, it is made entirely of German silver. There is a tuning slide but no water key.

It is thought that this tuba was made about 1847 or earlier. All of the other instruments with Paine valves are marked with the 1848 patent date and have a later, improved version of the valve. Paine is known to have played tuba in the American Brass Band so it is logical that he might have designed one early in his career.

The Rhode Island Historical Society has three instruments by Paine donated to the Society by former members of the American Brass Band of Providence. The first of these to be considered is an alto horn in A-flat (air column 144 cm) made of German silver and equipped with a telescopic tuning device (figure 7). A nameplate attached to the bell is inscribed "Thos. D. Paine & Co. Woonsocket, R.I. Patented 1848," (figure 7A). The instrument has three Paine valves of the improved design with valve tubes exiting from one end of the rotor instead of the sides. These valves are smaller and more compact than the earlier model yet they use the early type of string linkage. They are ar-

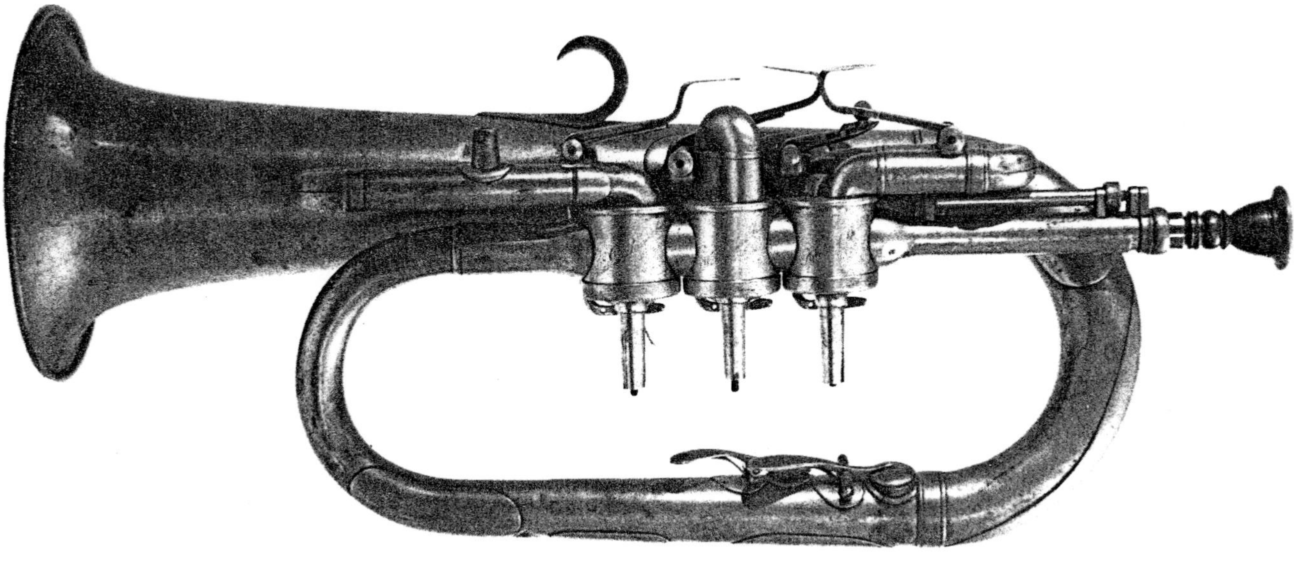

Fig. 8 – Valve and key bugle in e-flat by Thomas D. Paine, Woonsocket, circa 1850. German silver. Three Paine improved valves and two teys. (Photos courtesy of Central Missouri State University, Warrensburg)

ranged with the shortest valve first. The workmanship is very neat and the finish is smooth and shiny.

A German silver valve and key bugle in e-flat (figure 8), in the Don Essig Collection at Central Missouri State University, Warrensburg, has three of Paine's improved valves and two key bugle-type keys. There is a telescopic tuning device and the valve levers are arranged so that the shortest valve is played by the first finger. Several makers attempted a combination of valves and keys on the same instrument early in the 1850s and in 1853 some of these attempts were mentioned in *Dodworth's Brass Band School.*[8] Players switching from key bugle to valve bugle were reluctant to give up some of the advantages of the old key bugle. The keys on this instrument are primarily for written *a''* and *b*-flat'' (above the treble staff). This instrument has a rolled bell edge like modern instruments instead of the wide metal band or bell garland usually used at that date to reinforce the bell edge. The inscription on the valves reads "Thos. D. Paine & Co. Woonsocket, R.I. Patented 1848."

The collection of Alfred F. Wood of Westerly, Rhode Island, contains a cornet in C inscribed "Thos. D. Paine & Co. Woonsocket R.I. Patented 1848." (See figure 9) This instrument was probably a later model, for it shows several improvements. The three Paine valves are all of the later type and are arranged with the whole-step valve first as on modern instruments. The bore of the instrument expands more slowly and never reaches the large conical proportions of Paine's other brasses. It is, therefore, a cornet and not a bugle. This reduction in the conical expansion also allows a tuning slide to be fitted without materially disturbing the bore. The valve lever arrangement is similar to that on the Rhode Island Historical Society's alto horn. The workmanship is neat and although the instrument is heavily tarnished, it appears to have been well finished. Valves one and three have removable slides, but there are no water keys.

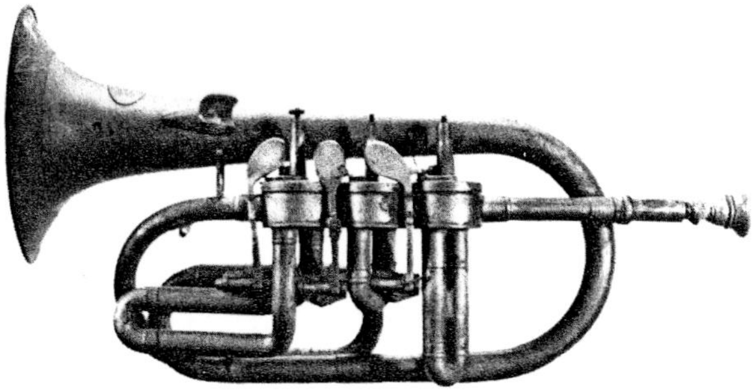

Fig. 9 – Cornet in C by Thomas D. Paine, Woonsocket, circa 1855. German silver. Three Paine valves. (Photo courtesy of Alfred Wood, Westerly, Rhode Island)

Another instrument found at the Rhode Island Historical Society is a key bugle in e-flat with 12 keys (figure 10). It is made of German silver, like most of Paine's instruments, but is one of only a few key bugles known of this material. Almost everything about this bugle matches the work of E.G. Wright. Its dimensions, bell garland, heart-

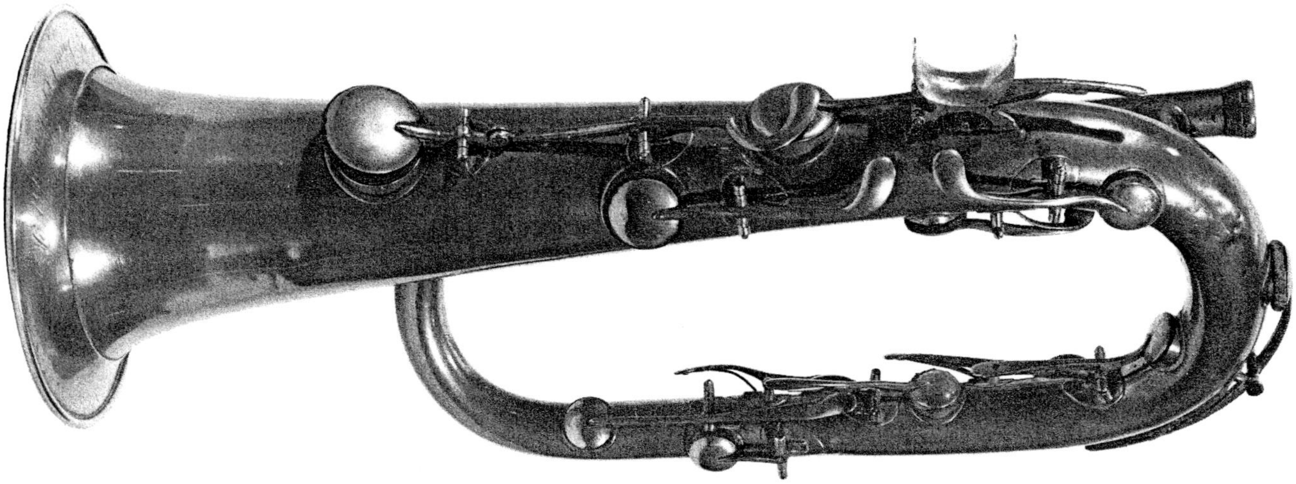

Fig. 10 – Key bugle in e-flat by Thomas D. Paine, Woonsocket, circa 1845. German silver. Twelve keys. (Photo courtesy of the Rhode Island Historical Society, Providence)

shaped foot-plates, key mounts, key shape, and the decorative saddle rest for the middle finger of the right hand are almost identical to those on bugles by Wright. One small difference is noticeable where the 12th key is mounted. On a Wright bugle the hole for key 12 (farthest from the bell) would be in line with the hole for key 11, and its foot-plate and key mount would be between holes 11 and 12. (See figure 26) Here hole 12 is above hole 11 and its key mount is very awkwardly squeezed against hole 11. Some of the foot-plate is cut away and the key is placed far to one side of the mount. It is likely that Paine worked with Wright in Boston in 1840, but even if he did not, Wright's bugles were very popular and much copied. Paine evidently used a Wright instrument as a model, but for some reason changed the placement of the last key. It is possible, too, that the last key was a later addition. This instrument is signed "Thos. D. Paine & Co. Woonsocket, R.I." and was made about 1842. It is missing its tuning shank and mouth-piece.

Also at the Rhode Island Historical Society is a tuba or bass in C (air column 240 cm) with six improved Paine valves (figure 11). It is of German silver, has a telescopic tuning device, and is very carefully finished. "Thos. D. Paine & Co. Pat. 1848 Woonsocket, R.I." is inscribed on a plate attached to the bell. The three valves for the left hand lower the pitch a half-step, a step-and-a-half, and a fifth, respectively. Those for the right hand are in the usual proportions except that the shortest valve is first. The left-hand valves have the early version of string linkage as found on most of Paine's instruments, but the right-hand valves have the string action in its improved version wrapping

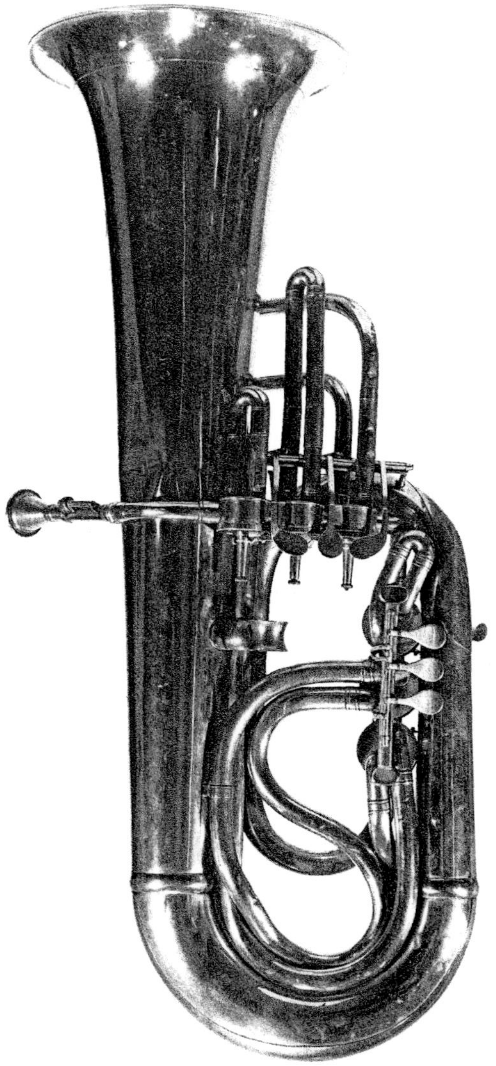

Fig. 11 – Tuba in C by Thomas D. Paine, Woonsocket, circa 1853. German silver. Six Paine valves. Figure 4 shows these valves in close-up. (Photo courtesy of the Rhode Island Historical Society, Providence)

once around the collar on the rotor shaft (figure 4). This version of the string linkage was widely used by many makers over the next century.

One other instrument by Paine, also a tuba or bass in C (air column 242 cm), is in the collection of Dr. G. Norman Eddy, Cambridge, Massachusetts. This instrument is made entirely of brass and has common type rotary valves turned by the improved string action. Its proportions, however, are those of the Paine instruments and it has the same narrow rounded ferrules at its main joints as are found on the tubas at the Henry Ford Museum and the Rhode Island Historical Society. The inscription around the bell garland is "Thos. D. Paine & Co. Woonsocket, R.I. Patented 1848."

Paine's early success quickly faded as other makers entered the brass instrument trade in the early 1850s. Better proportioned instruments were designed and the common European rotary valve proved superior to Paine's three passage rotaries. String linkage was promptly adopted by virtually all of the other makers and when Paine & Co. offered no further improvements, they lost business and gradually ceased production. In 1857 John Paine sold the shop and lot to Thomas[9] and retired. Thomas returned to watchmaking, and Emery continued mostly as a performer and teacher.[10] The census of 1860 listed John Paine, 75, without occupation; Emery A. Paine, 38, musical instruments; And Thomas D. Paine, 47, watchmaker.

A valve bugle signed "Emery A. Paine, Woonsocket, Rhode Island. Patented 1848" recently found by Eric Selch of New York City probably dates from the late 1850s after the brothers and their father separated. It is almost identical to the bugle shown in Figure 8, but has a bell garland instead of a rolled bell edge and is without the additional key bugle keys.

T. D. Paine's first wife, Perly, died March 2, 1861, and on April 8, 1863, he married Mary Arnold, a widow of some means; she purchased about half of John Paine's property.[11]

The latter part of T. D. Paine's long life is of little interest to music historians except that he began making and repairing violins. This new venture probably developed as a hobby, for until 1885 Paine's principal activity seems to have been watch and clock making and repairing. City directory and business directory listings, as well as census reports and two marriage records, from 1858 to 1885 all list his profession as jeweler and watchmaker.

Beginning in 1885, when he was 73, Paine began listing himself in city and business directories as a violin maker and repairer. A book published the following year listing prominent businessmen of Rhode Island included the following in an article on Paine:

> Mr. Paine has long been known as a watchmaker, a manufacturer of violins, and as a general mechanic . . . as a manufacturer of violins, his patronage is widely extended. These instruments are made in the best style of the art, and recommend their own merits on trial.[12]

Fig. 12 – Violin by Thomas D. Paine, Woonsocket, No. 11, 1856. (Photo courtesy of Herbert K. Goodkind, Larchmont, New York)

A check of violin dealers and collectors turned up only one violin by Paine (figure 12) in a private collection in Rhode Island. It is labeled "Thomas D. Paine Musical Instrument Maker No. 11 Woonsocket R.I. 1856." Even though it was made early in Paine's violin making career, it shows fine workmanship, good proportions and beautiful materials. The maple back is of one piece with especially nice figuration. The scroll is distinctive and its tone quality is good although not exceptional. If the quality of Paine's violins improved from number 11 to number 134 some may have been fine instruments. An inventory of Paine's estate after his death included fifteen violins, but these have not been traced. Paine's third wife, Ida Eleanor Darling Paine died in the State Hospital for the Insane at Cranston, Rhode Island, February 21, 1912; and his only child, Grace Darling Paine, born in 1890, has not been traced later than 1916 when she claimed to be Paine's only descendant.[13]

Paine's obituary in *The Evening Reporter*, Monday, June 3, 1895, included the following about his watch and clock work:

> For many years the tall form and patriarchal beard of the deceased were often seen on local streets. His fame as a watch and clock maker and repairer was not confined to the state or town where he resided. He was known all over southern New England as an expert in that line. For many years he had charge of the clocks in the Providence and Woonsocket stations of the Providence and Worcester road, and at one time the railroad men thought no man could repair a watch or clock as well as 'Old Tom Paine' of Woonsocket . . . He made many clocks, and one of them, a tall affair of polished wood showing the hours, adorns a room at his late home.

An obituary in the *Woonsocket Evening Call* mentioned his violin making.

> Mr. Paine's fame as a violin maker was as great as his skill in repairing watches. During his long life he has made 134 violins, finishing the last one the week previous to his death. His violins were much sought after by musicians, and $200 has been paid for one made by Thomas Paine. John Stromberg, a well known orchestra leader of New York, played on one of Mr. Paine's violins.

Thomas Dudley Paine's working life spanned seventy-three years and several careers. He will be remembered, however, only for a few minor improvements in brass instrument design made during a period of about ten years from 1847 to 1857. Throughout the latter half of the nineteenth century most American brass instruments had rotary valves with Paine's string linkage. Even after the turn of the century when piston valves became more popular, many instruments, particularly French horns, continued to use Paine's invention. Today the idea is still in use, although better mechanical linkage seems to be replacing string linkage on many new instruments.

J. Lathrop Allen

From the 1830's until the Civil War, Boston was the leading American city in the production of brass musical instruments. Among the foremost of a dozen or so makers who flourished in Boston during this period was J. Lathrop Allen. His work presents a fairly accurate picture of the development of brass instruments in the United States from about 1838 until 1870, and his inventions contributed to the rapid improvement of these instruments.

Brass instruments at this point in history had only recently acquired new mechanical means of altering their pitch. Two new systems were in use—side holes with covers manipulated by keys (key bugles, key trumpets, ophicleides) and valves which added short lengths of tubing to the instruments (cornets, valve bugles and trumpets, valve horns, etc.). Although valve instruments were gaining in favor, it was by no means certain at this time that the key brass would eventually be discarded. J. Lathrop Allen made some key brasses, but was known mainly for his valve inventions and fine valve instruments. E. G. Wright, on the other hand,

Throughout the third quarter of the 19th century brass instruments with Allen valves were very popular in the United States. These valves combined the advantages of short quick action, smooth quiet operation, simple construction and easy maintenance. Their only flaw, the distortion of the windway within the valve, prevented them from continued success; but until improvements were made in other types of valves, they ranked among the best available. The influence of the inventor can be seen in identical valves made by several contemporary makers.

Joseph Lathrop Allen was born in Holland, Massachusetts, just a short distance from Sturbridge on September 24, 1815.[14] His parents, Captain Ezra Allen and Lucena Fuller Allen, were the second generation of Allens to live on a homestead near Holland.[15] About 1838, J. Lathrop set up shop as a music instrument maker in Sturbridge. It is not known where he learned the trade since his father was a farmer, his grandfather a carpenter, and no known music instrument makers worked in that area. In 1839, he married Phebe S. Partridge of Brimfield, Massachusetts.[16]

Allen's earliest known instrument (about 1839) is a nine-key bugle in e-flat signed "Allen & Co., Sturbridge, Mass." in the collections of Old Sturbridge Village, Massachusetts (figure 13). It is made of copper with brass keys and trim and is very similar to other American bugles of that day. This type of bugle was a popular solo instrument and the leading voice in American bands of the period. The most common European bugle at this time was a larger instrument in B-flat with about seven keys.

The earliest valve instrument by Allen is a trumpet in B-flat with three Vienna twin-piston valves made after he moved to Boston in 1842. It is signed "J. Lathrop Allen, Maker, No. 16 Court Square, Boston" and is in the collection of Dr. Thomas R. Beveridge of Rolla, Missouri (figure 14). It has its half-step valve first, pointing up the fact that even by that time the now standard arrangement of valves (whole-step, half-step, step-and-a-half) had not been entirely decided upon, at least in America. Compared to later instruments, the bore, which is about 9 mm., is quite narrow and the arrangement of tubing and valves makes it appear upside down when held in playing position.

After moving to Boston in 1842, Allen can be traced quite easily through city directories and advertisements. The following appeared in notices placed regularly in *The American Journal of Music and Musical Visitor*, by B. A. Burditt, Boston music dealer:

> "B. A. Burditt is agent for the sale of Allen & Co's. celebrated brass instruments viz: valve post horns, valve trumpets, trombones, ophiclydes, and bugles; and has a supply on hand cheap for cash."[17]

The ad continued from July 1844 to February 1845, nearly eight months. Although it implies a certain amount of success, Allen could not have been completely satisfied, for shortly after this date, he moved his business to Norwich, Connecticut.

A left-handed tenor horn with slides for C or B-flat, from the period at Norwich, is in the collections of the Henry Ford Museum, Dearborn, Michigan (figure 15). It is similar in construction to the trumpet made in Boston. The same Vienna twin-piston valves are used, this time in standard order, and the instrument is shaped in a narrow upright form. It is made of copper and brass and is signed "Allen & Co., Norwich, Conn."

If business was slower than expected in Boston,

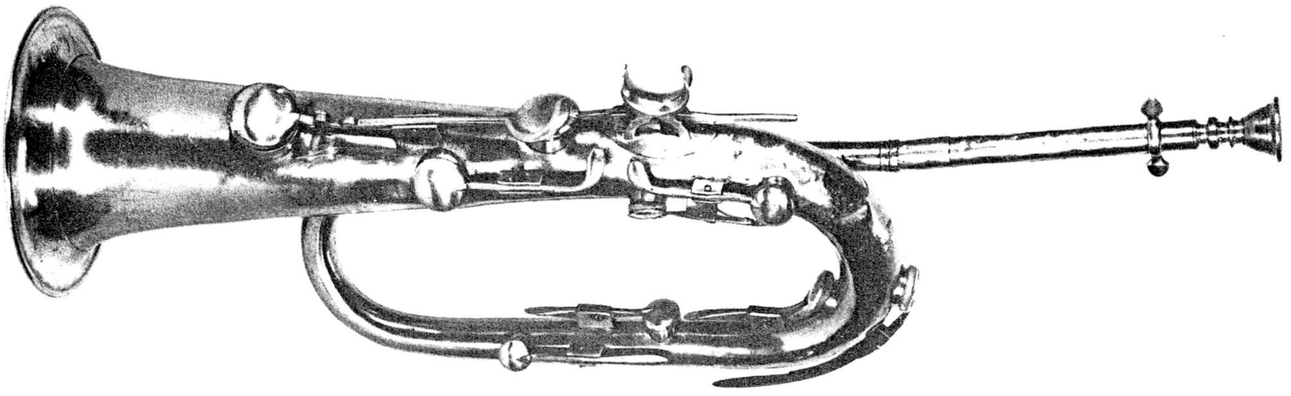

Fig. 13A, B – Key bugle in e-flat. "Allen & Co., Sturbridge, Mass.," circa 1839. (Photos courtesy of the Old Sturbridge Village, Sturbridge, Massachusetts)

Fig. 14 – Trumpet in B-flat. "J. Lathrop Allen, Maker, No. 16 Court Square, Boston," circa 1843. (Photo courtesy of Dr. Thomas R. Beveridge, Rolla, Missouri)

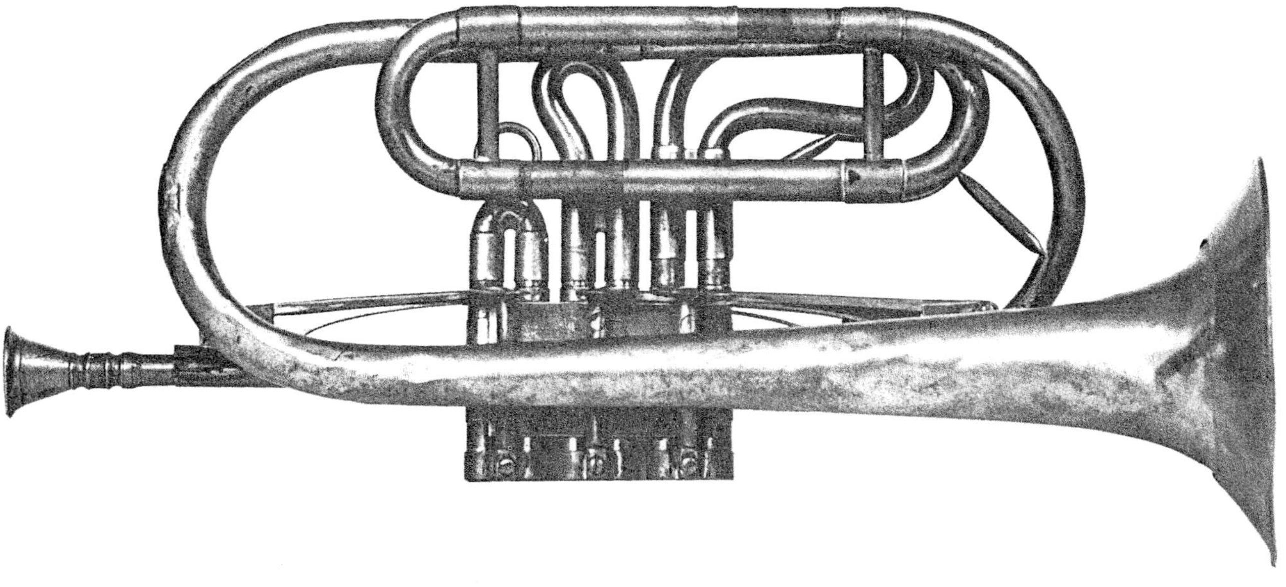

Fig. 15 – Tenor horn in B-flat or C. "Allen & Co., Norwich, Conn.," circa 1846. (Photo courtesy of the Henry Ford Museum, Dearborn, Michigan)

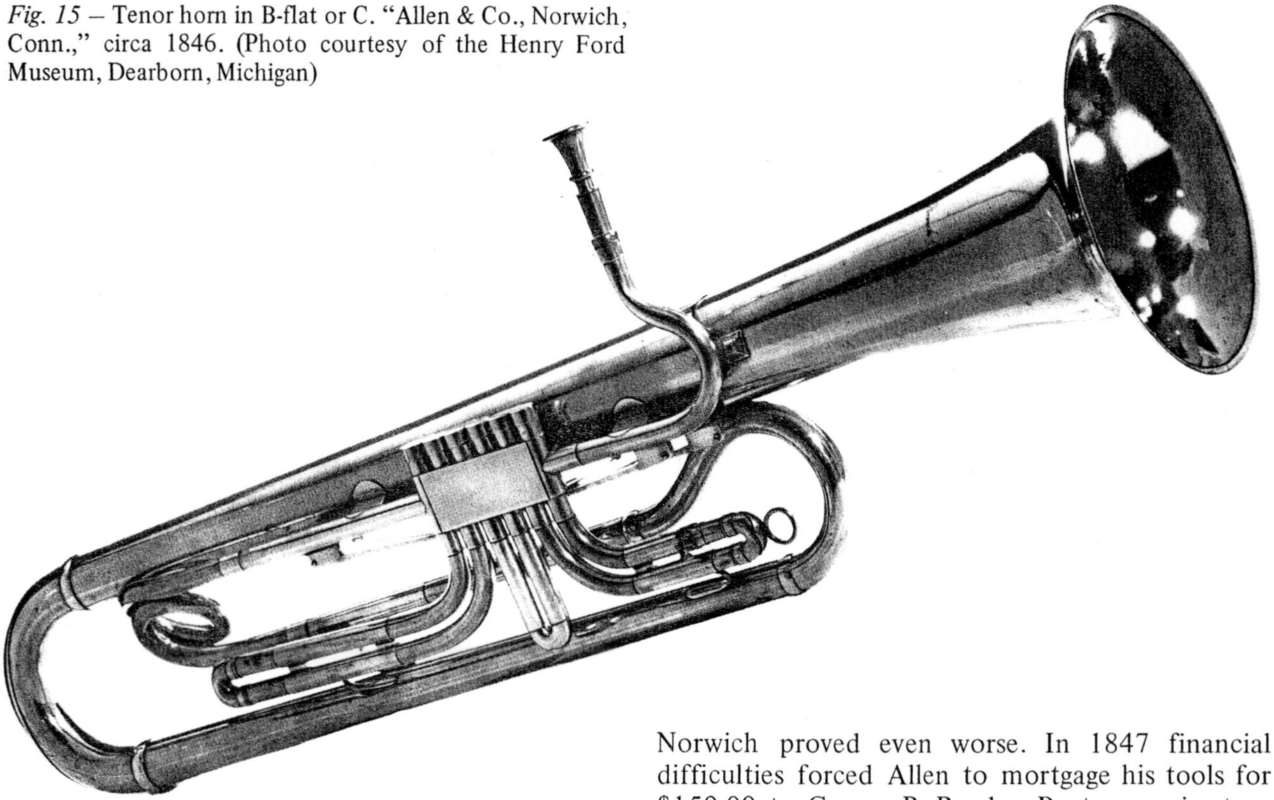

FRANKLIN SQUARE, FROM EAST MAIN STREET.

Fig. 16 – View of Norwich, Connecticut, circa 1875. Lithograph reproduced from *The Leading Business Men of Norwich and Vicinity*. Boston: Mercantile Publishing Co., 1890, p. 19. (Photo courtesy of the Otis Library, Norwich, Connecticut)

Norwich proved even worse. In 1847 financial difficulties forced Allen to mortgage his tools for $150.00 to George P. Reed, a Boston music store owner. The following list was found in the deed records of New London County, Connecticut. It has been rearranged to place related tools together and larger machines first.

- 1 Lathe and fixtures, including wheel and pulleys
- 2 Lots, milling tools including all that go in lathe
- 1 Polishing lathe
- 1 Bellows and forge tools
- 1 Anvil and block
- 3 Swage blocks
- 1 Draw bench and dies
- 1 Grindstone and frame, 1 oilstone
- 4 Tin benches
- 1 Rolling mill
- 16 Curved rods, 21 mandrils, stakes for forming, and wood patterns
- 2 Plotting irons, 2 try squares, 1 bevel (for measuring angles)
- 1 Pair dividers

Files, saws, scratcher broaches (reaming tools), shears, draw shave, reamers, chisels, hammer, planes, mallets, a screw wrench, an ax, and an auger.[18]

Allen ran a retail music shop as well as his manufacturing business at 2 Chapman's Block, Franklin Square (figure 16). The following advertisements which ran in the Norwich *Courier* for several months during the winter of 1847-48, give a good picture of his activities as well as an amusing glimpse of one of his sidelines.

PIANO FORTES.—The subscriber, Agent for the sale of **HALLETT, CUMSTON & ALLEN's** Piano Fortes, (late Hallett, Davis, & Co., Boston) has just received a splendid black walnut Instrument, of superior tone and finish, with Harp Pedal and all the late improvements, which he would be happy to show to those about purchasing. He would also state, for the benefit of those whom it may concern, that as his time is exclusively devoted to the Manufacture and Sale of Musical Instruments and Musical Merchandize only, (and having no Agents to pay commissions for recommending an instrument, as most dealers through the country have,) he is enabled to sell at much lower prices than others.

☞ A word to the wise is sufficient.

J. LATHROP ALLEN,
No 2 Chapman's Block, Norwich, Ct

Nov 27, 1847 eowtf39.

Seraphines and Melodions.

THE subscriber has been appointed agent for the sale of FARLEY & PEARSON's celebrated SERAPHINES and MELODIONS, which for quality of one and finish cannot be equalled in this country. Just received, a supply which the public are requested to call and examine.

ALSO—Where may be had all kinds of Musical Instruments, Sheet Music, Instruction Books for all kinds of Instruments, Violin, Violincello, Double Bass, Guitar Strings, &c &c.

Musical Instruments Manufactured and Repaired, and all work done when *promised*.

ALSO—Agent for the sale of Doct. Fontanie's Balm of Thousand Flowers, warranted to prevent the Hair from Falling Off - the best article for the Toilet ever sold. **J. LATHROP ALLEN,**
No. 2 Champman's Block.

Norwich, sept 9 tawtf28

Fig. 17 – Norwich *Courier* advertisements of November 27, 1847 and September 9, 1847.

Fig. 18 – Allen's flat windway valves; and an Allen rotor compared with the common rotary valve rotor. (Photo courtesy of the Henry Ford Museum, Dearborn, Michigan)

Fig. 19 – Overshoulder tenor horn in B-flat. "Manufactured by J. Lathrop Allen, 17 Harvard Pl., Boston," circa 1855. (Photo courtesy of Fred Benkovic, Wauwatosa, Wisconsin)

Sometime during the years from 1847 to 1852, Allen designed the very efficient rotary valve for which he is remembered. It was based on the same principles as the Vienna rotary valve dating from 1832, but had several remarkable features. The rotors of Allen's valves were much smaller in diameter, but longer; and the passages within them were shaped like a flat oval, half the thickness of the round tube in their narrowest dimension and more than twice the size of the round tube in their widest dimension. The tubing remained round as usual until two or three centimeters before each valve. At this point the passage was gradually flattened, much like a round heating duct as it

Fig. 20 – Overshoulder cornet in a-flat. "Made by J. Lathrop Allen, 17 Harvard Pl., Boston, for H. B. Dodworth, N.Y.," circa 1855. (Photo courtesy of the Henry Ford Museum, Dearborn, Michigan)

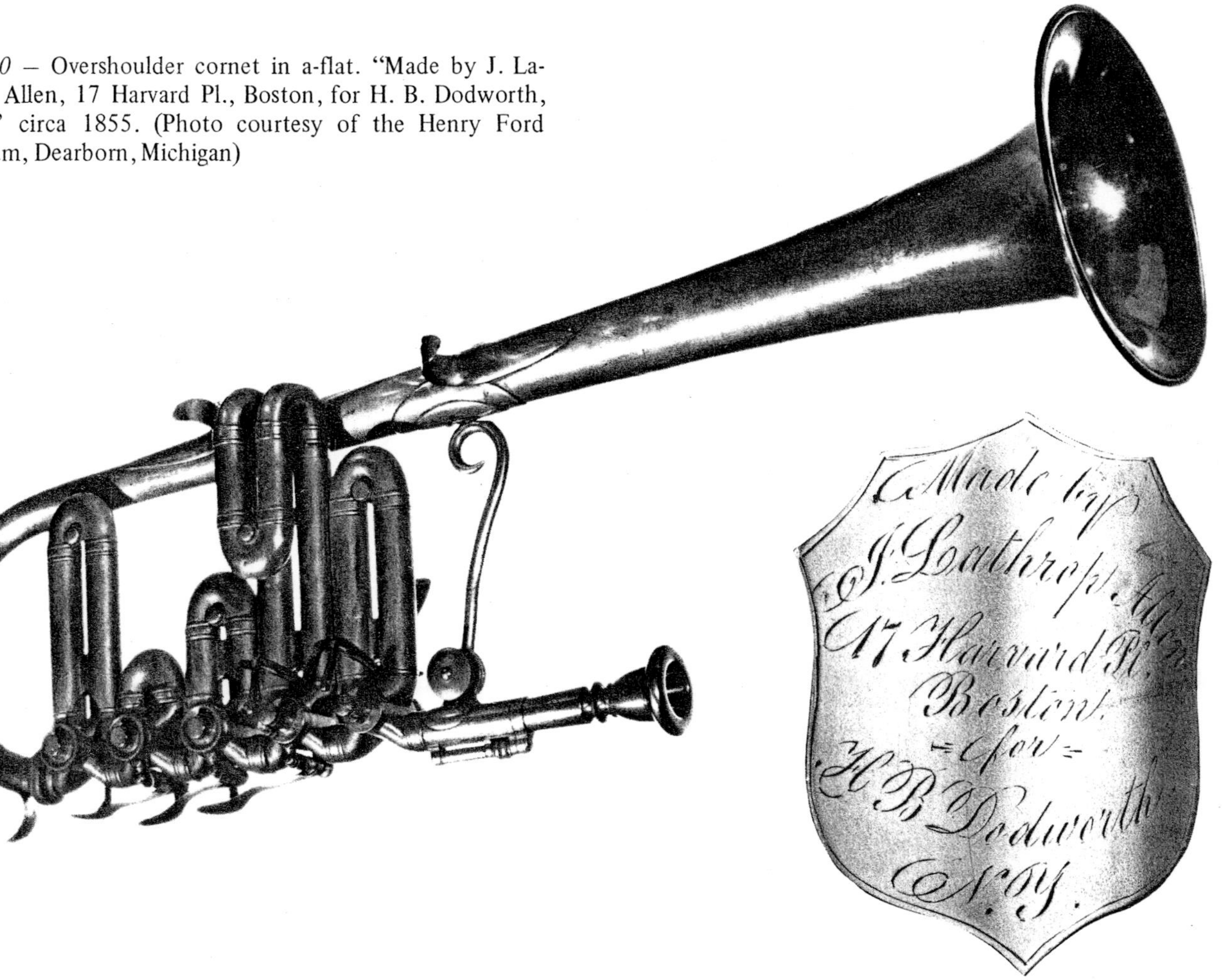

approaches a rectangular wall duct. The size of the oval passage was designed to retain the exact capacity of the preceding round passage. This bit of engineering resulted in a valve rotor that was easier to turn and stop, simply because it was considerably smaller in diameter.

Allen then used string linkage to turn his valves, an idea first tried about 1848 by Thomas D. Paine of Woonsocket, Rhode Island. String linkage provided an advantage in leverage as well as in smooth, quiet operation. His final improvement was to place the stopping corks on the end of the rotor so that they were enclosed by the cap—a simpler and quieter arrangement. These valves had a very light, quick touch and rapidly became popular among many American players (figure 18).

Allen returned to Boston in 1852, and was soon producing instruments incorporating his improved valves. One of the earliest of these known today is a tenor horn in the collection of Fred Benkovic, Wauwatosa, Wisconsin (figure 19). It is made entirely of brass and is larger in bore than the earlier Norwich tenor. The improved proportions and valves make it a much more playable instrument.

About 1855 Allen made an unusual overshoulder cornet for Harvey B. Dodworth, a well-known New York bandmaster. It is in the key of a-flat and has five Allen valves. In addition to the usual three for the right hand, it has two for the left hand lowering the instrument to f or e-flat. It is made of German silver and is in the collections of the Henry Ford Museum (figure 20). With the endorsement of this well-known band leader and soloist, Allen was soon able to expand his business.

Although J. Lathrop Allen never entered instruments in fairs or exhibits himself, his products were occasionally entered by music stores and dealers. The following reports by the judging committee of the Fifteenth Exhibition of the Ohio Mechanics' Institute, held in Cincinnati, Ohio in 1857, give an idea of Allen's stature at this time.

One cornet-a-piston made by J. Lathrop Allen of

Fig. 21 – View of the west side of Washington Street, Boston showing the side street Harvard Place. Reproduced from *Gleason's Pictorial Drawing Room Companion*, May 14, 1853, p. 313.

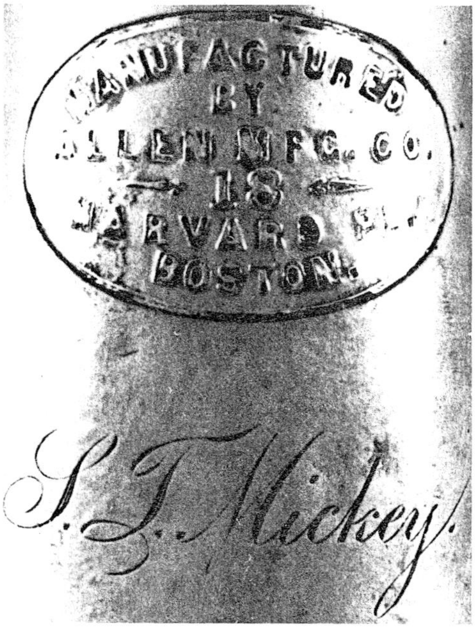

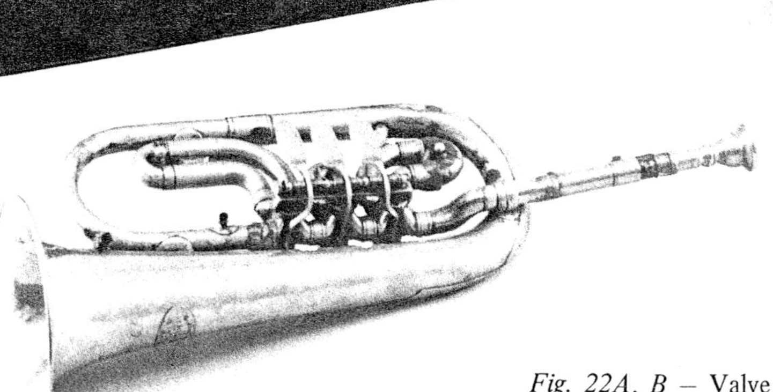

Fig. 22A, B – Valve bugle in e-flat with case. "Manufactured by Allen Mfg. Co., Harvard Pl., Boston," circa 1859. (Photos courtesy of Old Salem Inc., Winston-Salem, North Carolina)

Fig. 23 – Overshoulder tuba in E-flat. "Allen & Hall, 331 Wash. St., Boston," circa 1861. (Photo courtesy of the Henry Ford Museum, Dearborn, Michigan)

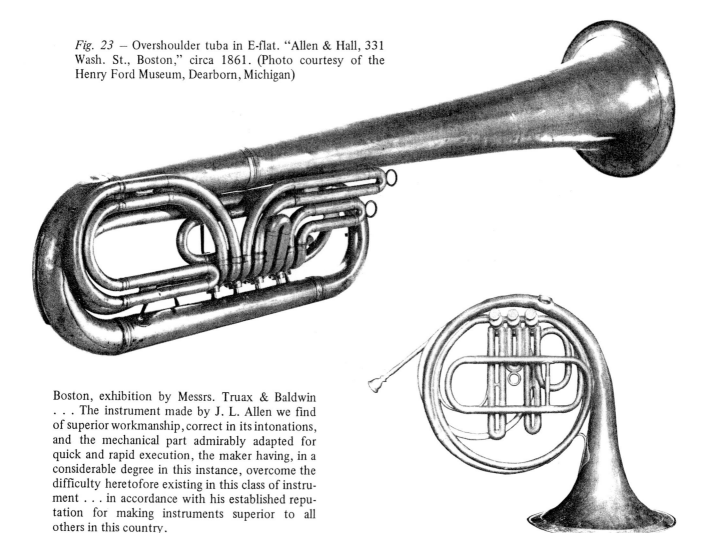

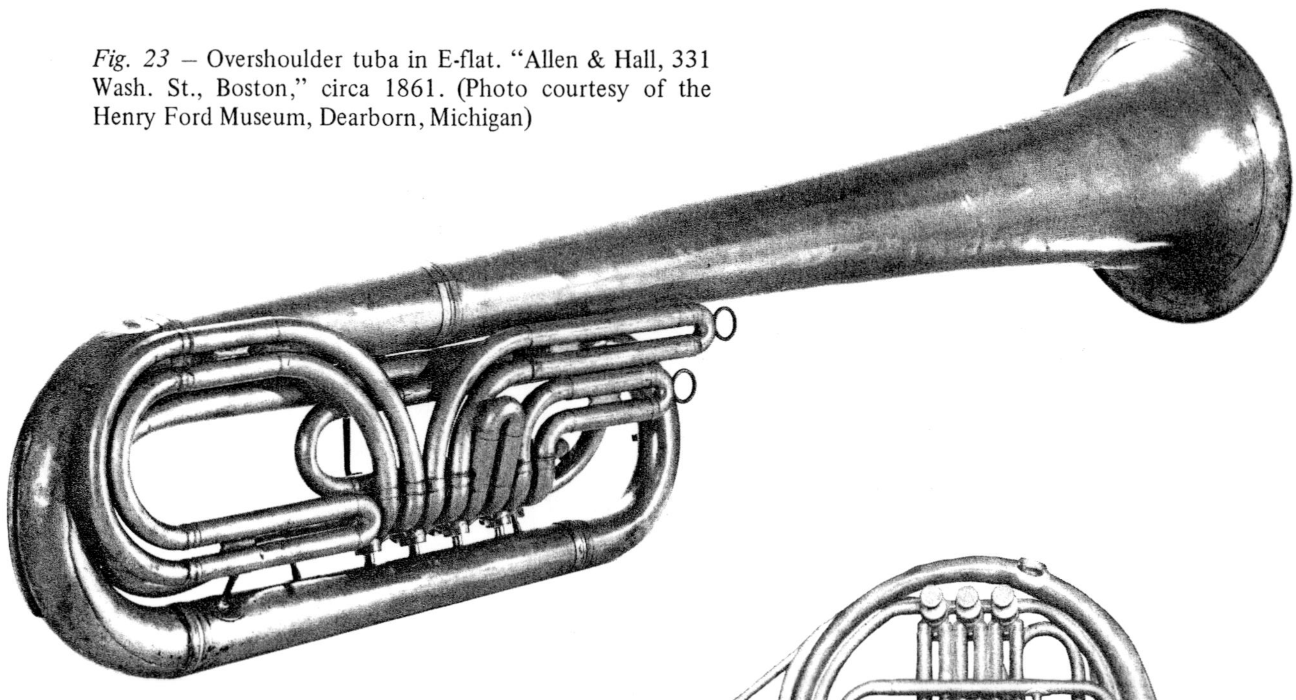

Fig. 24 – French horn. "J. L. Allen, Maker, N.Y.," circa 1870. (Photo courtesy of the Metropolitan Museum of Art, New York City)

> Boston, exhibition by Messrs. Truax & Baldwin . . . The instrument made by J. L. Allen we find of superior workmanship, correct in its intonations, and the mechanical part admirably adapted for quick and rapid execution, the maker having, in a considerable degree in this instance, overcome the difficulty heretofore existing in this class of instrument . . . in accordance with his established reputation for making instruments superior to all others in this country.
>
> One e-flat alto (silver), made by J. Lathrop Allen, and exhibited by Messrs. Truax & Baldwin, we find of superior workmanship and excellent tone, the mechanical part being perfect in its construction, giving utility and beauty to the instrument, and is but another evidence of the well deserved reputation of the makers.[19]

From 1852 until 1860 Allen's shop in Boston was located at the end of Harvard Place, a little dead-end street off Washington between Milk and School Streets. Harvard Place was also the address of E. G. Wright from 1861 to 1863, and at various times housed many of the other Boston makers of musical instruments (figure 21). It has now been entirely built over and not even one of those street signs of historic Boston remains to show where it once was located.

Allen's success during the 1850s enabled him to buy a house at 33 Porter Street in 1857, evidently the first real estate he had ever owned.[20] He continued to expand his business, calling it the Allen Manufacturing Co. (See figure 22)

During 1852 and 1853, Allen had a partner named Benjamin F. Richardson. Those working for him in 1858 included Anton, Erhardt, and Franz Huttl; August Doelling; Joseph Koestler; L. F. Hartman; and Henry Esbach. For a short time at the beginning of the Civil War, David C. Hall, a Boston band leader and key bugle soloist, was a partner in Allen's business. A number of instruments of this period are signed Allen and Hall. (See figure 23) Both Hall and Benjamin F. Richardson continued to make Allen valves for their instruments after leaving Allen's shop. Many of these valves are found on instruments signed by B. F. Richardson, D. C. Hall, and Hall and Quinby.

After the Civil War, Allen moved his business to New York and worked there until 1872. Thereafter

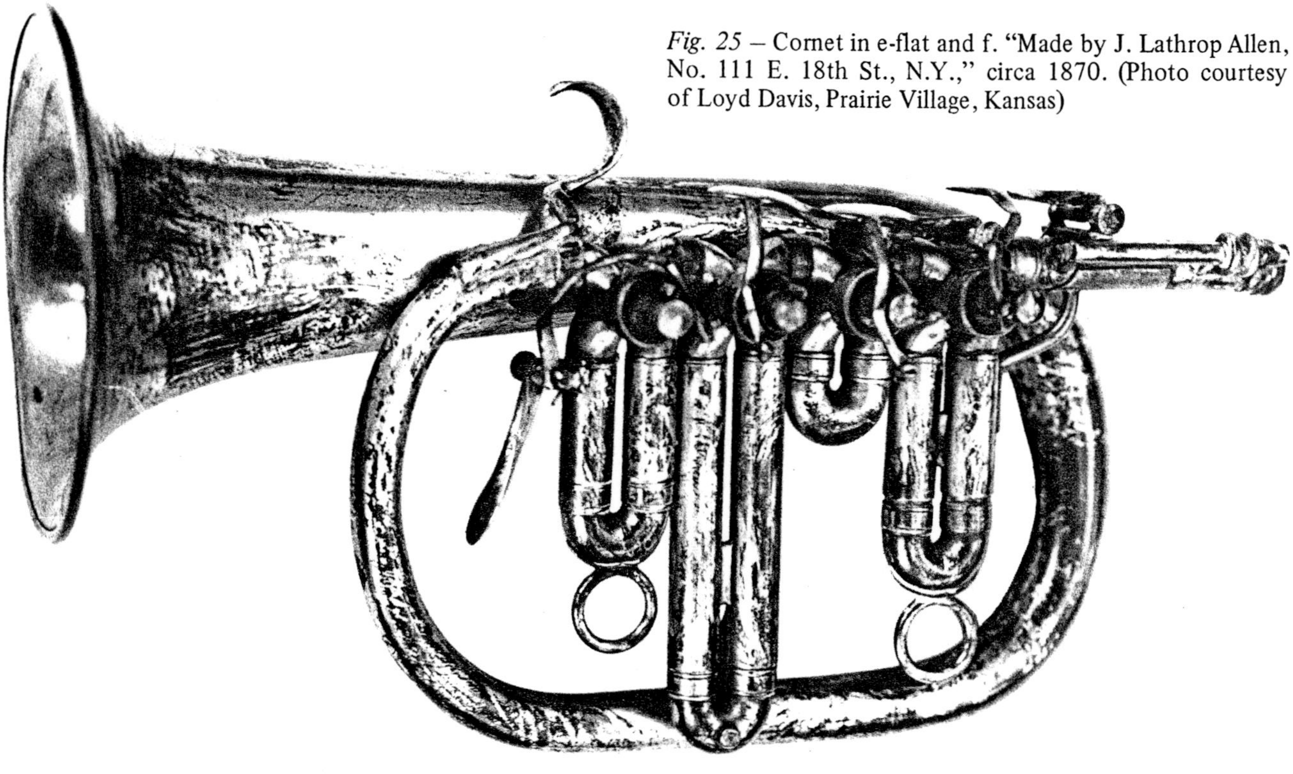

Fig. 25 – Cornet in e-flat and f. "Made by J. Lathrop Allen, No. 111 E. 18th St., N.Y.," circa 1870. (Photo courtesy of Loyd Davis, Prairie Village, Kansas)

he probably made few instruments. In 1877 he is known to have been back in Boston; from 1880 to 1897, he appears again in the New York directories. Two instruments made by Allen when he was working in New York are a French horn with Berlin valves, now in the Metropolitan Museum of Art collections in New York City (figure 24), and a cornet belonging to Loyd Davis of Prairie Village, Kansas (figure 25). Genealogical records indicate that he lived to be over ninety although his death date is not known.[21]

Allen was important to American brass making not only because of his unique valves and influence among several other craftsmen, but also because of the fine workmanship displayed on all of the instruments he produced.

E. G. Wright

By the 1850s, the American brass band had come of age. From its beginnings, about 1835, until this time, it had developed from miscellaneous groups of ophicleides, trombones, key bugles, post horns and whatever other brasses were handy to fairly homogeneous groups of mostly valve instruments. The saxhorns of Adolph Sax were the pattern for many kinds of brasses made in sets and similar in tone from soprano to bass. These sets, which formed the basis for many bands, were advertised in the papers and magazines with instrumentations for every size band from six to seventeen. A typical advertisement by Firth & Pond in the *Musical World & N.Y. Musical Times*, December 4, 1852, showed proper instrumentation for bands of eight, ten and twelve brass instruments:

Band of Eight	Band of Ten	Band of Twelve
2 E-flat Soprano	3 E-flat Soprano	3 E-flat Soprano
2 E-flat Tenor	2 E-flat Tenor	2 E-flat Tenor
1 E-flat Alto	1 E-flat Alto	2 E-flat Alto
1 B-flat Baritone	1 B-flat Baritone	1 B-flat Baritone
1 B-flat Bass	1 B-flat Bass	2 B-flat Bass (four valve)
1 E-flat Contra	2 E-flat Contrabass	2 E-flat Contrabass

The one holdover from the earlier instrumentation, however, was the solo e-flat key bugle, Many of the band leaders were specialists on this instrument and had developed considerable skill and technique in performing intricate solos with their bands. These accomplishments were not easily given up and often resulted in the use of a solo key bugle with the valve brasses through the 1850s.

Brass bands were popular and enjoyed a lot of community interest and loyalty. Band leaders, of course, were often the focus of this interest and it is not surprising to discover that they were presented with tokens of community esteem upon occasions of retirement after long service, outstanding accomplishment, assuming the leadership of a band, or the formation of a new band.

An appropriate gift for an outstanding key bugle soloist was, of course, a fine key bugle, and examination of surviving presentation instruments and published accounts of such gifts reveals the most celebrated maker of key bugles to have been E. G. Wright of Boston. Of thirteen presentation instruments where the maker is known, twelve were made by Wright. He was clearly the outstanding bugle maker of the period and his instruments were in demand not only for their fine playing quality but also for the excellent workmanship and decorative detail lavished on them.

Most of Wright's best e-flat bugles were made of silver. One of the finest of these is a twelve key instrument in the collections of the Henry Ford Museum (figure 26). It is one of the few examples known at present which is complete with its original form fitting case. A decorative gold plate on the bell is engraved "E. G. Wright, Maker, No. 115 Court Street, Boston." Other engravings cover the bell garland, mouth-pipe, and even the tuning shank.

An even finer example in materials and workmanship is another instrument from the collections of the Henry Ford Museum which was "Presented to D. C. Hall Esq. by the members of the Lowell Brass Band, April 15, 1850." It is made entirely of gold and is profusely decorated with engravings on the bell and mouth-pipe, around each tone hole, and even on reinforcing strips protecting vulnerable areas of the tubing (figure 27). Each axle where the keys pivot has a tiny sleeve protecting the gold key from wear. A gold-plated duplicate of this instrument made by D. C. Hall is also at the Henry Ford Museum. Ten other silver bugles and one of silver with gold keys by Wright are known at present. Each instrument is decorated differently and each reveals the maker as a master craftsman.

Elbridge G. Wright was born in Ashby, Massachusetts, March 1, 1811; one of at least eight children in the family of Elijah Wright and Levina Lawrence Wright.[22] Both were of solid pioneer families going back several generations in New England. Ashby is a small town in the northwestern tip of Middlesex County. The Wrights were a successful and well thought of farm family.

Wright may have worked at another trade for several years, since his first known attempts at music instruments making were in the late 1830s when he was already 27 or 28 years old. The earliest of his instruments found so far is an ophicleide of about 1839 in the Essig Collection, Warrensburg, Missouri, signed "E. G. Wright, Roxbury." Unfortunately this ophicleide is missing its mouth-pipe, but it appears to have been in C and has nine keys (figure 28). It is made of brass and has flat hole covers made for leather pads. Later ophicleides had as many as eleven keys with cupped hole covers for stuffed pads. Roxbury, although now a part of Boston, was a small neighboring town in the 1830s. Wright evidently worked there for a time before starting his shop in Boston

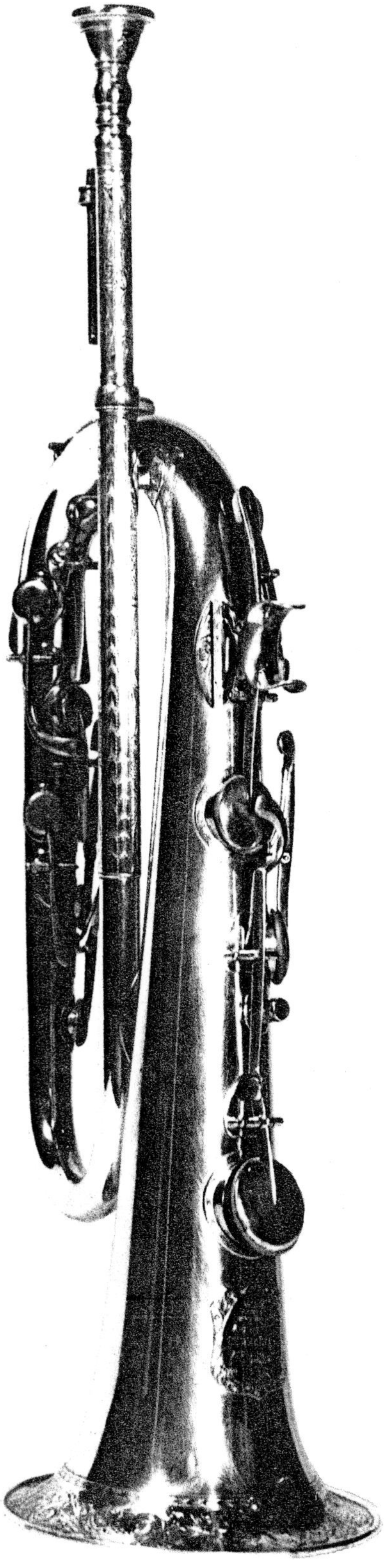

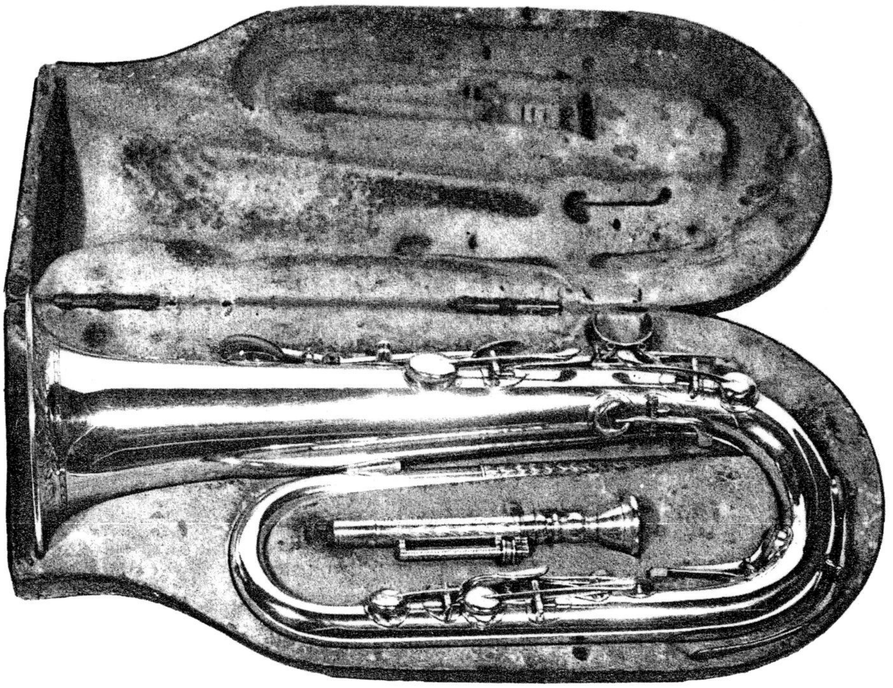

Fig. 26A, B – Key bugle in e-flat. "E. G. Wright, Maker, No. 115 Court St., Boston," circa 1850. (Photos courtesy of the Henry Ford Museum, Dearborn, Michigan)

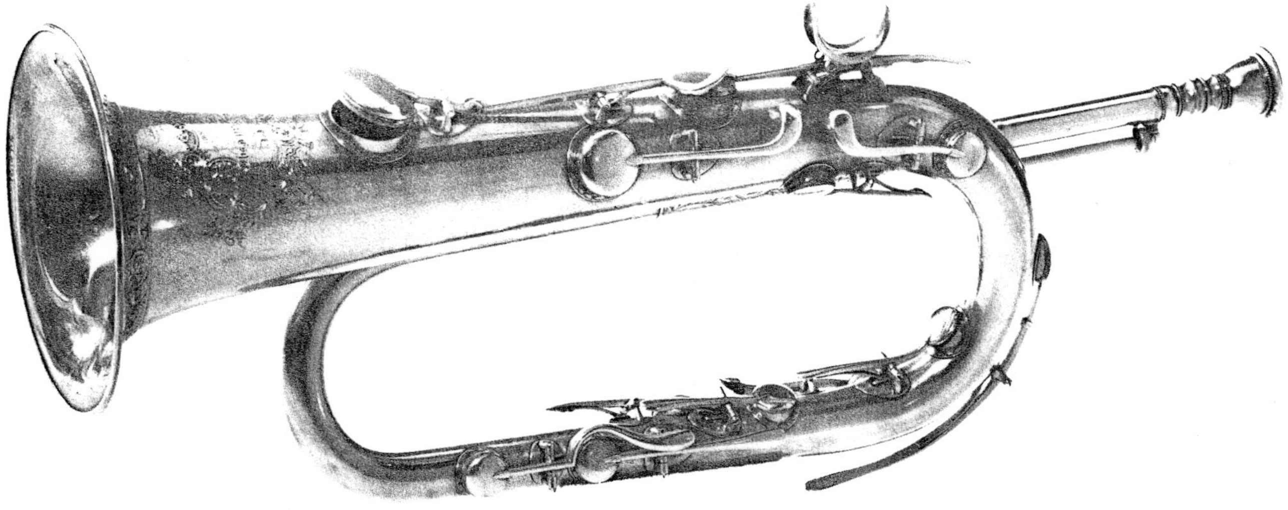

Fig. 27A, B, C – Key bugle in e-flat. "E. G. Wright, Boston," 1850. (Photos courtesy of the Henry Ford Museum, Dearborn, Michigan)

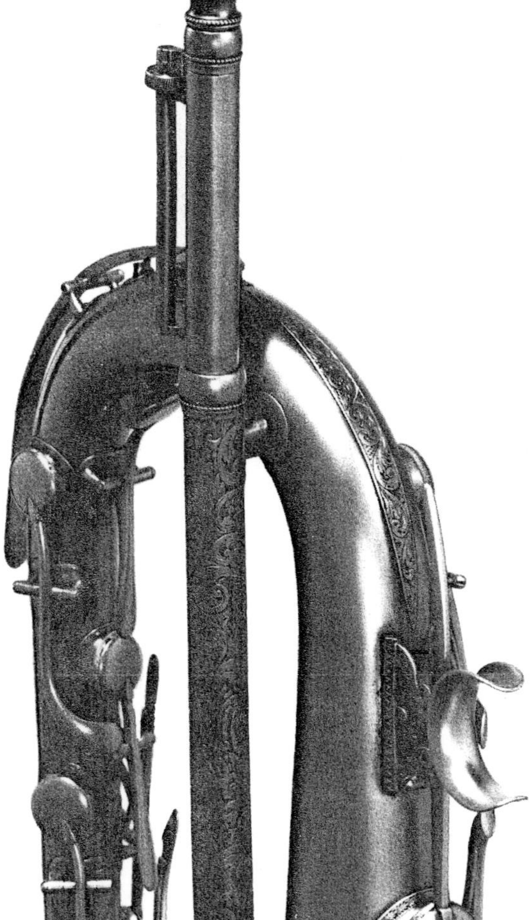

Fig. 28 – Ophicleide in B-flat. "E. G. Wright, Roxbury, Mass.," circa 1839. (Photo courtesy of the Central Missouri State University, Warrensburg, Missouri)

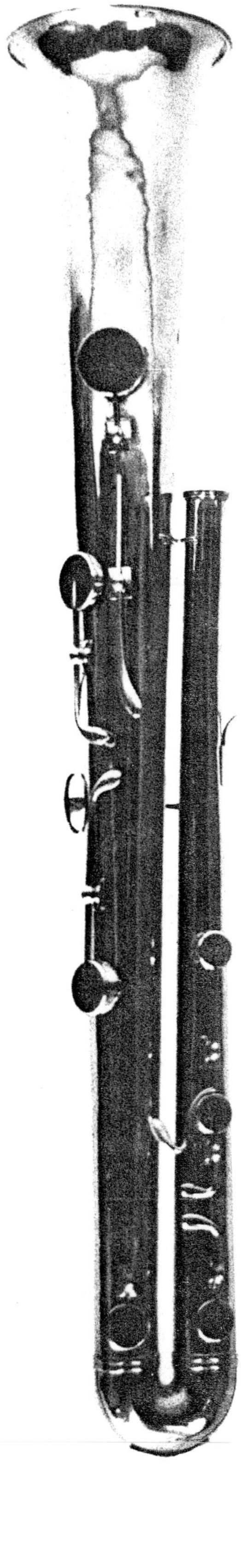

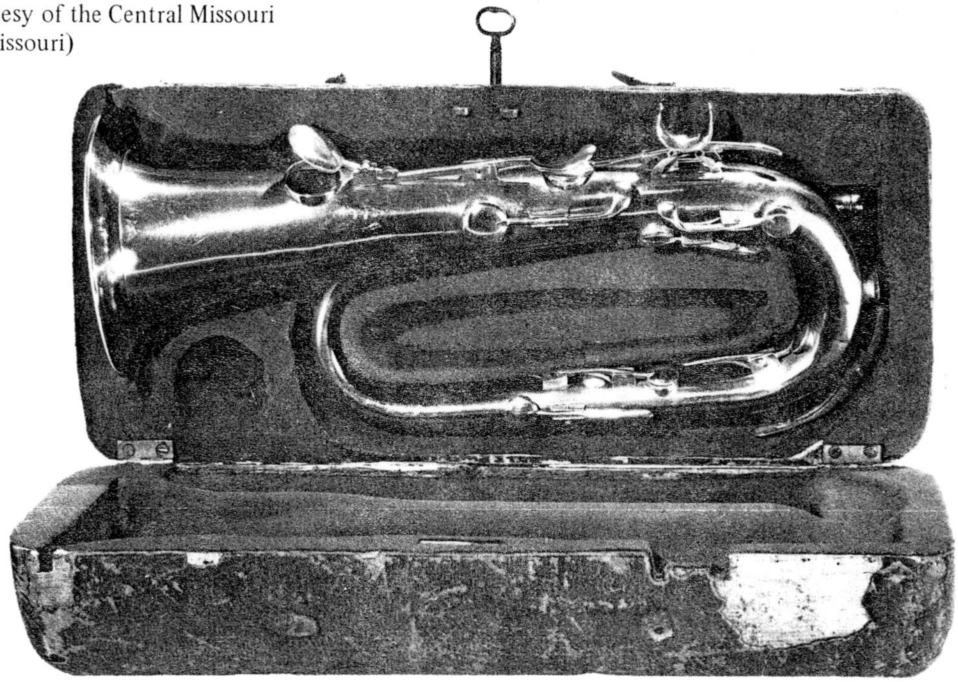

Fig. 29 – Key bugle in e-flat with case. "E. G. Wright, Maker, Boston," circa 1845. (Photo courtesy of the Henry Ford Museum, Dearborn, Michigan)

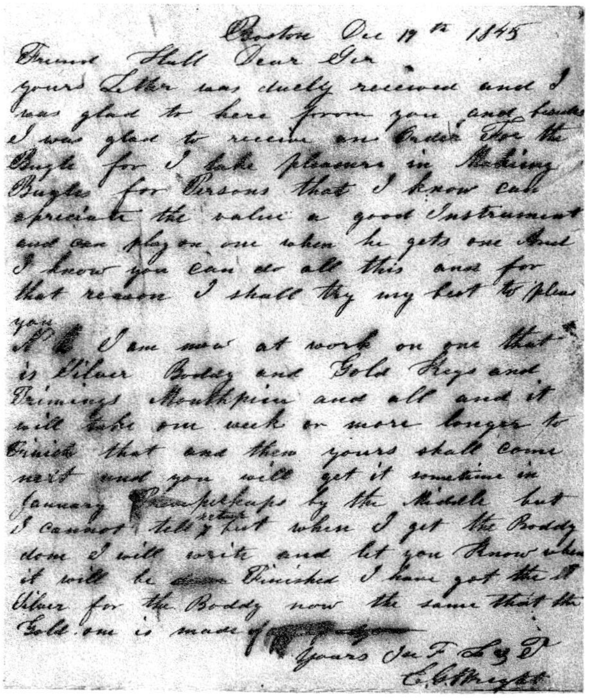

Boston Dec 17th 1845

Friend Hall Dear Sir

yours Letter was duely received and I was glad to here from you and beside I was glad to receive an Order for the Bugle for I take pleasure in Making Bugles for Persons that I know can apreciate the value a good Instrument and can play on one when he gets one And I know you can do all this and for that reason I shall try my best to please you

N B I am now at work on one that is Silver Boddy and Gold Keys and Trimings Mouthpiece and all and it will take one week or more longer to finish that and then yours shall come next and you will get it sometime in January perhaps by the Middle but I cannot tell exactly but when I get the Boddy done I will write and let you know when it will be finished I have got the Silver for the Boddy now the same that the Gold one is made of

Yours Ju F Lo Co D

E G Wright

Fig. 30 – Letter from E. G. Wright to D. C. Hall, December 17, 1845. (Photo courtesy of the Henry Ford Museum, Dearborn, Michigan)

BOSTON BRASS BAND.

Fig. 31 – Lithograph of the Boston Brass Band reproduced from *Gleason's Pictorial Drawing Room Companion*, August 9, 1851, p. 225.

in 1841. His son George C. was born in Roxbury in 1839.[23]

Wright's original Boston shop was on Bromfield Street and there he began to build a reputation for himself as an excellent craftsman and fine musician. His most popular products at first were e-flat key bugles made of copper with brass or sometimes German silver trim, and equipped with about nine keys. Typical of this type of instrument is a bugle in the Henry Ford Museum collection, made of copper with brass trim and ten keys with a form fitting case hollowed from a solid block of wood (figure 29).

About every three years the Massachusetts Charitable Mechanic Association held an exhibit in Boston displaying mechanical objects of every kind made by local and neighboring craftsmen. Wright entered a key trumpet in the exhibit of 1841 but the judges' only comment was that it "came in too late for examination."[24]

By the mid-1840s, Wright had begun to capture the market for fine solo key bugles. The following letters in the collections at the Henry Ford Museum are addressed to David C. Hall, then a rising young band leader and key bugle soloist who became a partner of J. Lathrop Allen in the 1860s (See figure 30). They were written during the winter of 1845 to 1846.

> Boston Dec 19th 1845
>
> Friend Hall Dear Sir,
>
> Your letter was duely received and I was glad to hear from you and besides I was glad to receive an Order for the Bugle for I take pleasure in Making Bugles for Persons that I know can appreciate the value a good instrument and can play on one when he gets one, And I know you can do all this and for that reason I shall try my best to please you. N.B. I am now at work on one that is Silver Body and Gold Keys and Trimmings Mouthpiece and all and it will take one week or more longer to Finish that and then yours shall come next and

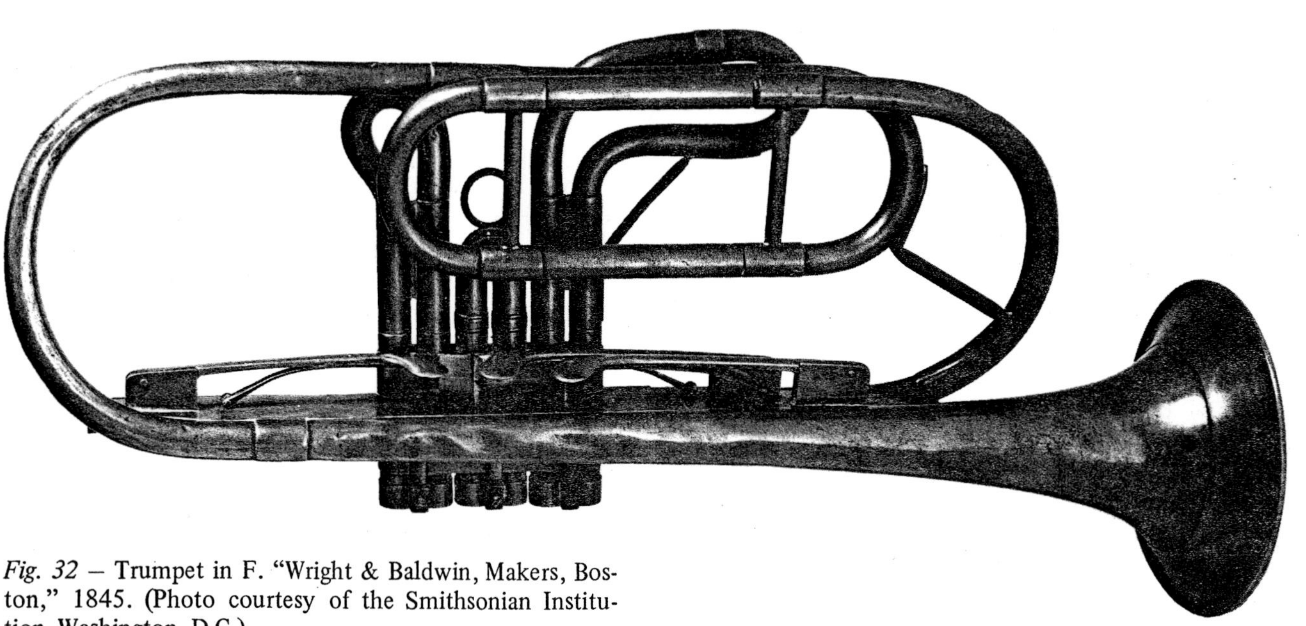

Fig. 32 – Trumpet in F. "Wright & Baldwin, Makers, Boston," 1845. (Photo courtesy of the Smithsonian Institution, Washington, D.C.)

you Will get it sometime in January perhaps by the Middle but I cannot tell certain but when I get the body done I will write and let you know when it will be Finished. I have got the Silver for the Body now, the same that the Gold one is made of.

Yours in F.S. & T.
E. G. Wright[25]

Boston, March 9th 1846

Friend Hall Dear Sir,

I suppose you are tired of waiting for your Bugle and I do not wonder, but I will explain the reason wh'is that I have been very busy this winter in Musick. I have played almost every night since the 1st of December and have not been able to do much in the shop, but the season is nearly over for Dancing now although we are engaged for two weeks ahead for every night, but I shall be at work on the Bugle all the time I can get and as I suppose you are not in a hurry for it, I shall take time to do it well and I will write you again soon and let you know it is getting along.

Yours & c E. G. Wright

N.B. Mr. B. Burdit and the remaining members of Kendall's Band wish me to write to you and request an answer immediately whether they can obtain your services to play the Es Bugle in the Band this season and likewise your Brother.

Yours & C. E. G. Wright[26]

Ned Kendall, famous New England key bugle player, formed the Boston Brass Band in 1835 and continued as its soloist and director until 1842. In 1846 Eben Flagg was the band's director and soloist[27]; and B. Burditt, a Boston music dealer and publisher, was evidently the manager (figure 31).

If E. G. Wright played for dancing almost every night from December 1, to March 9, and was booked for at least two more weeks, dancing appears to have been a very popular form of entertainment in Boston at that time and Wright's band must have been a favorite. The letters also indicate that Wright was a bit of a procrastinator which is borne out by his second attempt to exhibit at the Massachusetts Charitable Mechanic Association Fair. In 1847 he entered a silver Bugle with gold keys. Again, the judge reported only that "it was entered too late and taken away too soon for consideration."[28] If that was the same bugle he started in the winter of 1845-46, and which he was to finish before starting the silver one for Hall, there are probably some far more interesting letters between Hall & Wright that remain to be found.

From 1848 until 1852, Wright's shop was located at 115 Court Street in downtown Boston. In 1853 he moved a few doors away to 121 Court, and in 1856 he left the downtown area and moved in with Samuel Graves, another instrument maker, at 68 Albany. For a short time during 1858 and 1859, Wright worked in Lowell, Massachusetts. In 1860 he is listed in Boston again at 27 Portland, and in 1861 he moved to 18 Harvard Place, a popular area for music instrument shops.

Along with his key bugles E. G. Wright also continued to experiment with valve instruments. His first attempts in this field were made with twin-piston Vienna valves of the same type used by

Fig. 33 – Title page of "Shelton's Quick Step." New York: C. G. Christman, 1852. (Photo courtesy of the Newberry Library, Chicago, Illinois)

Allen. A trumpet in F with these valves is preserved in the Smithsonian Institution. It is signed "Wright & Baldwin makers, Boston," a partnership which existed only in 1845 (figure 32). Late in the 1840s the development of string linkage for rotary valves made possible a very smooth, quiet valve action. In the early 1850s Wright began making instruments with string operated Vienna rotary valves almost exactly like those found in French horns today. He also made some instruments with both keys and valves in an attempt to gain the advantages of both systems. This attempt was described by Allen Dodworth in his *Brass Band School* published in 1853.

> "Soprano cornets have lately been made in this country, combining the advantages of both valves and keys; they have three valves, like the ordinary cornet, with the addition of five keys for the upper notes; the one nearest the bell for the highest A-flat, that with the next for A, the second and third for B-flat, the third and fourth for B, and the fourth and fifth for C; this is a very great improvement as they combine the fullness of tone in the lower notes peculiar to valve instruments with the greater ease and facility of the upper notes which is peculiar to keyed instruments."[29]

An illustration on the front of a piece of sheet music published in 1852 shows James Shelton, Esq., band leader and bugle soloist, holding a valve and key bugle (figure 33). So far, the only existing instrument that exactly fits Dodworth's description is an overshoulder bugle in e-flat by E. G. Wright in the Henry Ford Museum collection (figure 34). It is made of German silver and is only a little smaller in bore than a key bugle. This instrument, together with Allen Dodworth's description, is not only interesting documentation of a transition between key and valve bugles but also reveals something else not previously known. The measurements from the bell of each of the keys coincides with those of the last five keys on American twelve-key bugles. The correct use of these

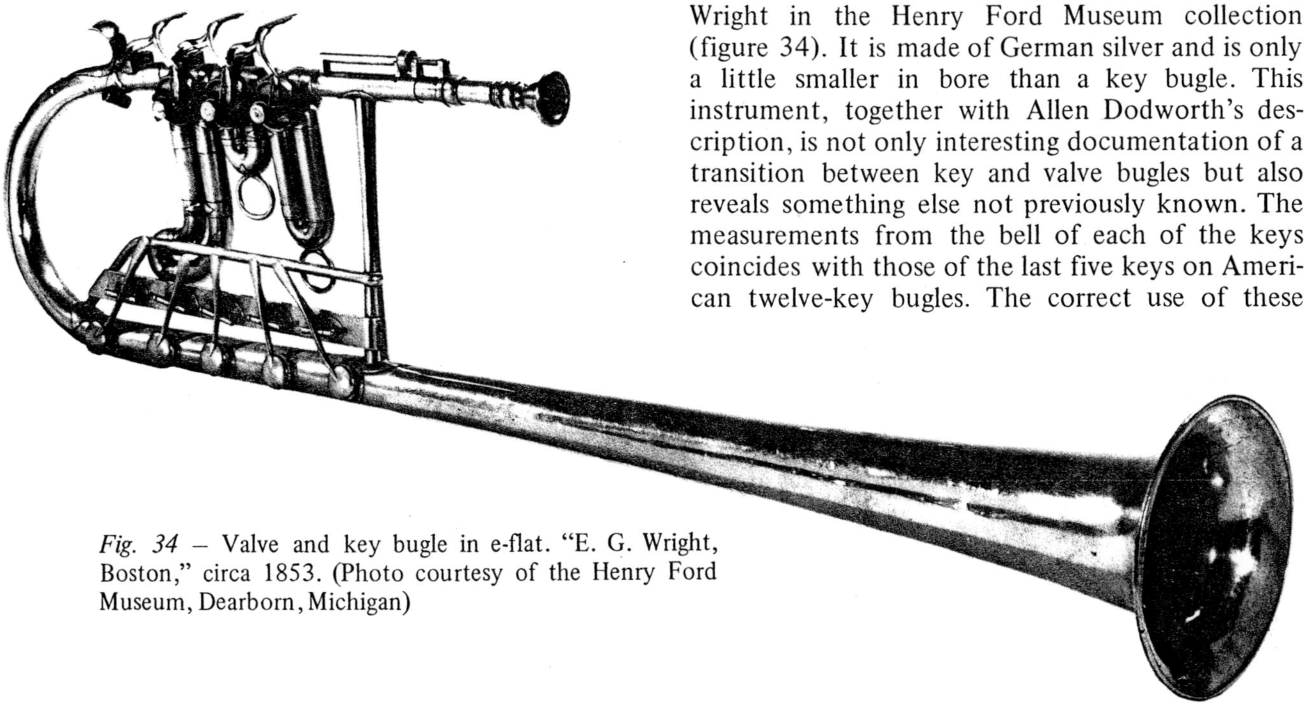

Fig. 34 – Valve and key bugle in e-flat. "E. G. Wright, Boston," circa 1853. (Photo courtesy of the Henry Ford Museum, Dearborn, Michigan)

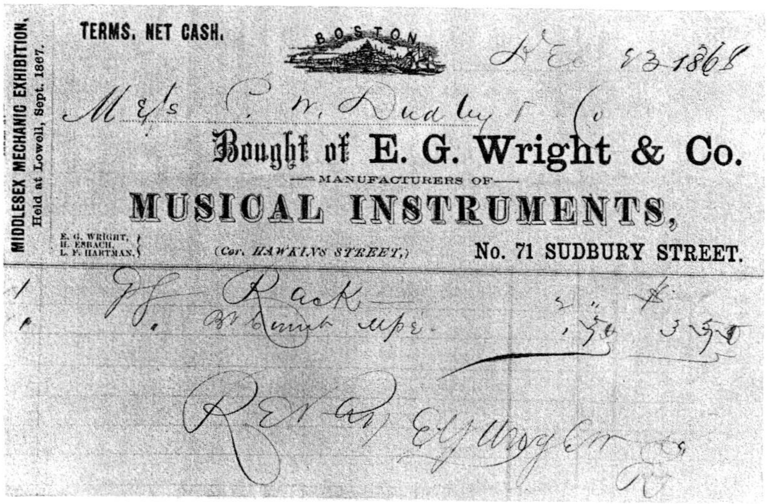

MIDDLESEX MECHANIC EXHIBITION, Held at Lowell, Sept. 1867.

TERMS, NET CASH.

BOSTON

Dec 23 1868

M ess C. W. Dudley & Co

Bought of E. G. Wright & Co.

MANUFACTURERS OF

MUSICAL INSTRUMENTS,

E. G. WRIGHT, H. ESBACH, L. F. HARTMAN.

(Cor. HAWKINS STREET,)

No. 71 SUDBURY STREET.

Fig. 35 – Receipt, 1868. From the Warshaw Collection. (Photo courtesy of the Smithsonian Institution, Washington, D.C.)

Fig. 36 – Two valve bugles in e-flat, one with case. "E. G. Wright, Boston," circa 1863. (Photo courtesy of the Smithsonian Institution, Washington, D.C.)

Fig. 37 – Circular cornet in B-flat. "Made by E. G. Wright, Boston," 1855. (Photo courtesy of the Henry Ford Museum, Dearborn, Michigan)

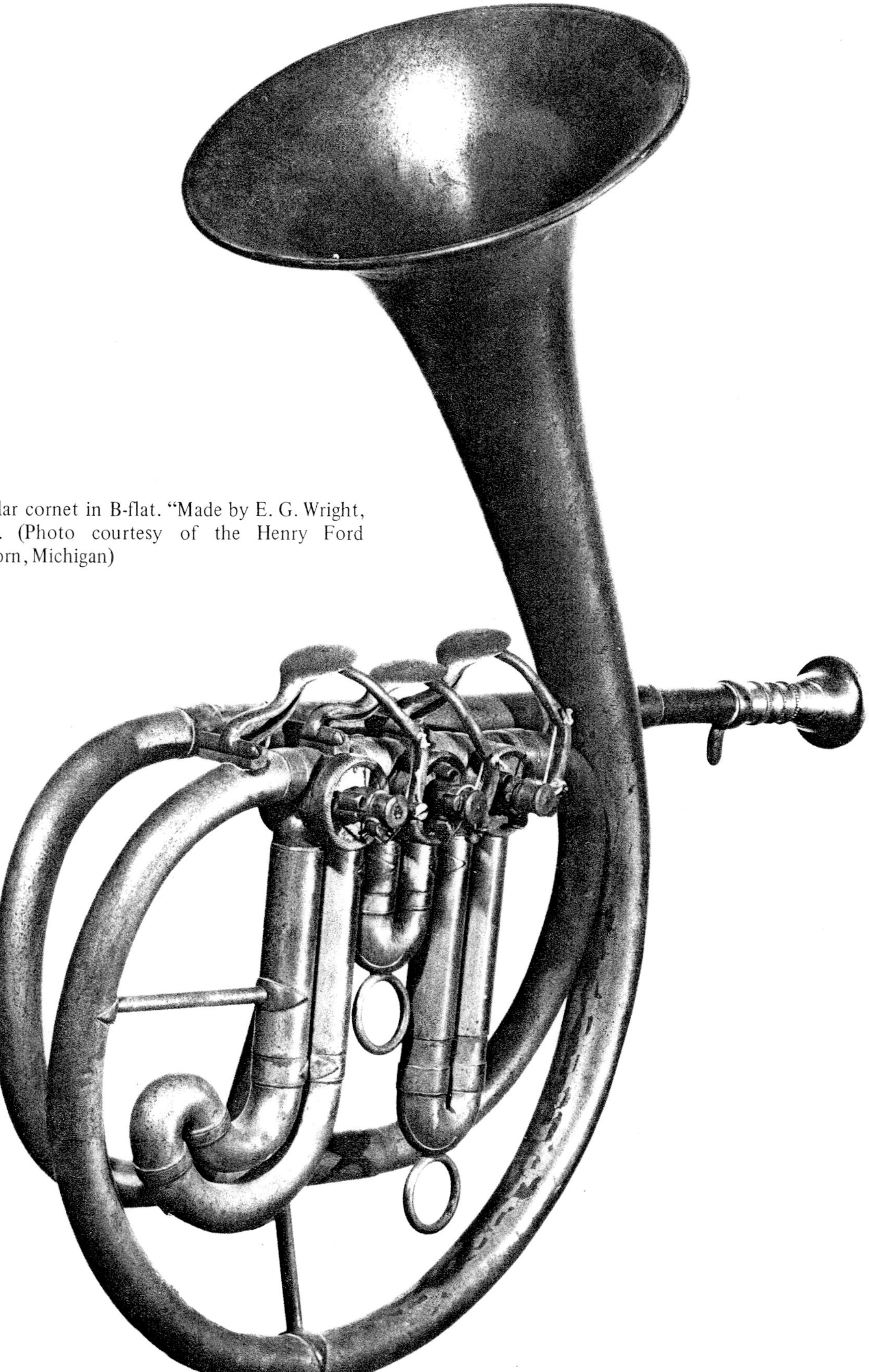

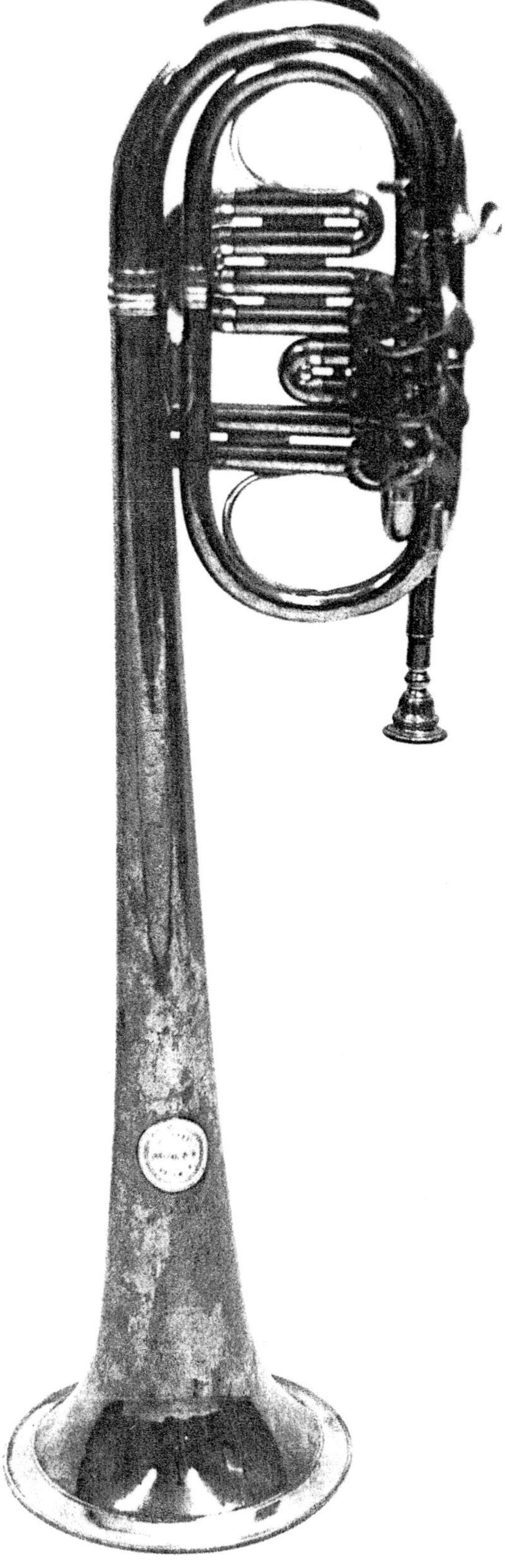

Fig. 38 – Overshoulder alto or tenor in E-flat. "E. G. Wright, Boston," circa 1865. (Photo courtesy of the Central Missouri State University, Warrensburg, Missouri)

Fig. 39 – Tenor or alto in E-flat. "Made by E. G. Wright & Co., Boston," circa 1865. (Photo courtesy of the Henry Ford Museum, Dearborn, Michigan)

Fig. 41 – Trade Card, 1867. From the Warshaw Collection. (Photo courtesy of the Smithsonian Institution, Washington, D.C.)

Fig. 40 – Overshoulder tuba in E-flat. "E. G. Wright, Maker, Boston," circa 1862. (Photo courtesy of The Stearns Collection, University of Michigan, Ann Arbor)

keys has been a matter of some inexact knowledge and the best fingering chart discovered so far covers only eleven keys. The use of these five keys evidently coincides with key bugle usage and, therefore, suggests the proper fingerings for the twelve key bugle.

The latter part of the productive life of E. G. Wright was devoted more and more to valve brasses. Beginning in 1864 E. G. Wright & Co. at 71 Sudbury, Boston turned out large quantities of these instruments in all sizes from E-flat bass to e-flat soprano (See figure 35). At this time Louis F. Hartman and Henry Esbach were working for Wright. In 1865 a full set of brasses entered in the Massachusetts Charitable Mechanic Association exhibit was awarded a first silver medal for superior tone and workmanship[30] (See figures 36-40).

For a time in 1867 the well-known bandmaster Patrick Gilmore was associated with Wright (See figures 41 and 42). About 1869 the firms of E. G. Wright & Co. and Samuel Graves & Co. merged to form the Boston Musical Instrument Manufactory. Wright withdrew in 1870, and worked with Hall & Quinby until his death in 1871.

Wright's finest key bugles are unsurpassed anywhere in the world. They are the pinnacle of development reached by that instrument. Although, because European makers had already turned their attention to valve instruments, this distinction may have been gained by default; it is still an accomplishment of considerable historical importance.

Fig. 42 – Tenor or alto in E-flat. "Made by Wright, Gilmore & Co., Boston, for W. & W. Mfg. Co.," 1867. (Photo courtesy of the Henry Ford Museum, Dearborn, Michigan)

Isaac Fiske

Throngs of people waited impatiently along a main street in Troy, New York, on a summer day in 1857. The anticipated event was a parade headed by the Troy Brass Band. In the crowd was Isaac Fiske from Worcester, Massachusetts, a maker of musical instruments with a special interest in seeing that Fiske's Cornet Band of Worcester (which used all Fiske instruments) was the finest band around. Band leaders vied with one another in those days not only in fine instruments and stirring music, but in who could field the tallest and most ornate drum major, the loudest drums, and the showiest uniforms. Fiske wanted to have a look at a young Scottish musician named Matthew Arbuckle whose fame had spread across the Berkshires.

Arbuckle had been chief piper in the kilted band of the Royal Scottish Regiment of Canada. He was a first class bagpipe player, cornet soloist, and drum major as well. Backers of the Troy Band had induced him to desert his regiment and settle in Troy. Now as he came into view, Isaac Fiske understood why. First he beat unerring time on the bass drum while flourishing two drumsticks in the air. Resuming his position as drum major, he would hurl his gold plated baton high overhead, catching it nimbly behind his back. Then, signaling for the band to play, he thrilled his listeners with high sweet solos on his e-flat cornet. Isaac Fiske was impressed, and when he went back to Worcester the next day, Matthew Arbuckle accompanied him.[31]

Arbuckle remained in Worcester for three years marching at the head of Fiske's band in the finest uniform Isaac Fiske could buy. In the end, however, he left in the same way he had come. The famous Gilmore Band visited Worcester in 1860 and Arbuckle put on such a show that Patrick Gilmore felt he had to have him for his band. In spite of a lawsuit by Fiske[32] Matthew Arbuckle joined Gilmore's Band. He later became their featured cornet soloist. Fiske had gotten his money's worth however. Arbuckle and the Fiske Band had put Isaac Fiske's instruments on the map just in time to reap a bountiful harvest from the rush of orders for band instruments during the Civil War.

Fiske, an excellent maker of brass instruments, remained active in Worcester, Massachusetts, from 1842 until he sold his business to C. G. Conn & Co. in 1887. He was inventive and sought to solve some of the problems of brass instruments of that day. Five United States patents were granted for his improvements and, although none of them are incorporated specifically in the instruments of today, they were steps in the right direction—toward lighter, faster-acting valves and clear unobstructed windways. These improvements and a substantial production place Fiske among the five foremost brass instrument makers of mid-19th century America.

Isaac Fiske was born in Holden, Massachusetts, September 17, 1820, son of Bezaleel and Mary (Rice) Fiske. The family sold their property in Holden in 1832[33] and moved to Worcester. It is not known how Isaac learned the trade of a music instrument maker, but he set up shop in 1842 at 77 Main. He was then twenty-two years old.

Worcester had a population of 7,497 in 1840 and an active musical life nourished by frequent concerts and three music stores. Pianos were sold by George A. Willard; Charles S. Ellis ran an umbrella and music store; and Leland & Putnam stood "ready to supply brass & wood musical instruments of every kind and quality."[34] Concerts during the years 1839-43 included visits by the Nicholson Flute Club of Boston; The Boston Brass Band, with Ned Kendall, famous bugle soloist; the Steyermark Family of instrumentalists; and a variety of local instrumental and vocal music performances. Although many musicians are named in these programs, Isaac Fiske does not appear among them. He is known to have been a cornetist but was evidently not a leading performer. As a maker of brass instruments though, he must have been heartened at the opportunity suggested by the following insertion in the *Worcester Palladium*, August 30, 1843:

> **Grand Musical Jubilee**
>
> The members of the Worcester Brass Band and other citizens of Worcester and vicinity . . . resolved to invite all the bands far and near to convene at Worcester of Friday, September 8, 1843 . . . up to 50 bands . . .

The results of this invitation went unreported in the paper, but something proved stimulating to the new musical instrument firm, for very shortly Fiske moved to larger quarters, expanded his stock of merchandise, and placed his first advertisement

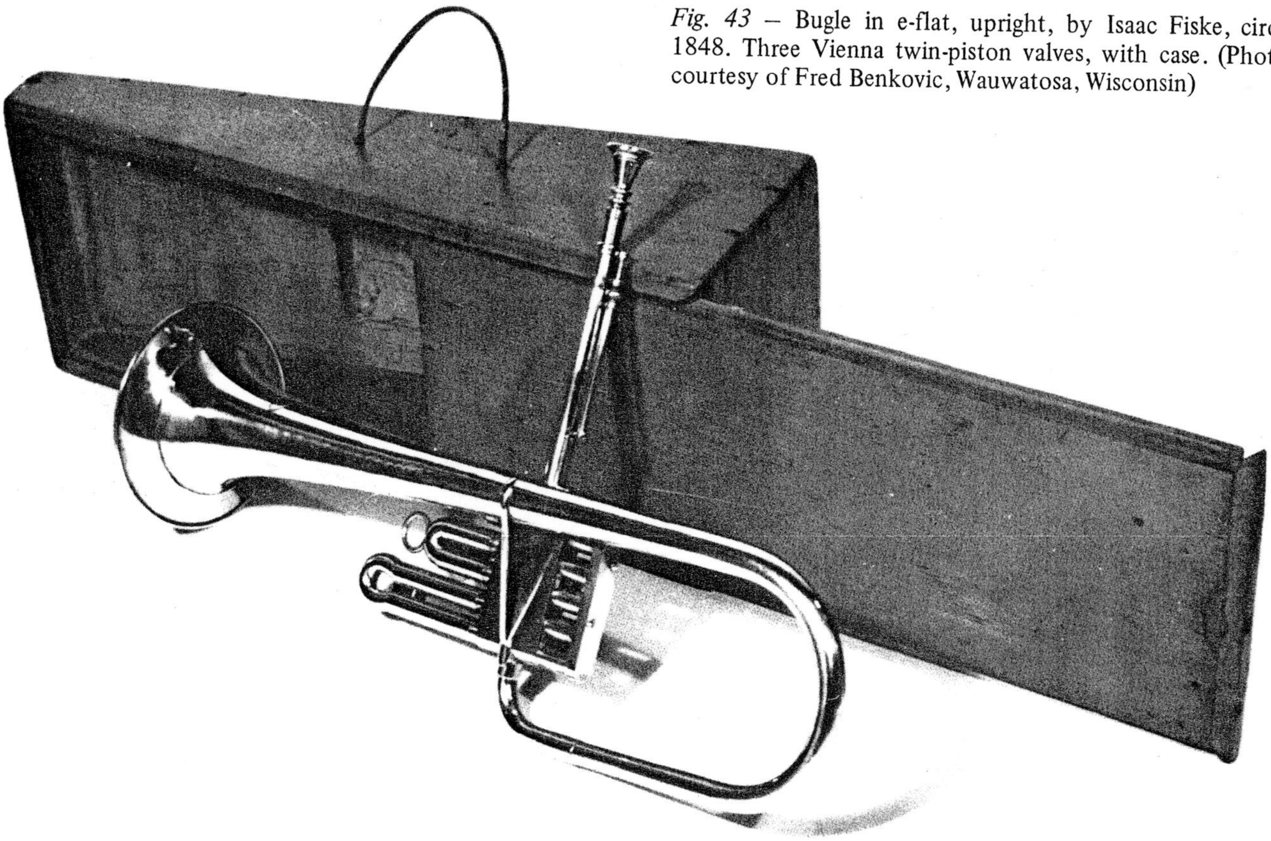

Fig. 43 – Bugle in e-flat, upright, by Isaac Fiske, circa 1848. Three Vienna twin-piston valves, with case. (Photo courtesy of Fred Benkovic, Wauwatosa, Wisconsin)

in the paper. The following appeared on September 27, 1843, in the same newspaper:

> **New Music Store and**
> **Instrument Manufactory**
>
> Isaac Fiske would respectfully inform his friends and former patrons that he has removed to 141 Main Street, where he will be happy to furnish them with musical instruments of all kinds, consisting in part of B. and E. flat bugles; B.C. and E. flat posthorns manufactured by himself and warranted correct; valve trumpets in the key of F; clarionetts; flutes; fifes; melodions & c.; opheclydes; bass and tenor trombones of a new style and superior to any ever yet offered to the public.
>
> A choice selection of new sheet music instruction books of all kinds–blank music paper, books and cards . . .

The times were ideal for brass instrument making. Since their introduction in 1835, brass bands had become the rage in New England. Two or three other New England makers had already begun producing brass instruments; Fiske was only a few years behind them and could profit by their experience. He may even have apprenticed with one of them. Certainly it was not far from Worcester to Winchester, New Hampshire, where Graves

& Co. operated, nor was it difficult to get to Boston where E. G. Wright worked.

Examples of the earliest products of Fiske's shop have not been found. The advertisement of 1843 mentions e-flat and B-flat bugles (probably key bugles) and B-flat, C and E-flat posthorns (probably the small circular type without valves). If he was not making valve instruments at that time, it was not long before he began.

From 1846 to 1850 Fiske shared a shop near the railroad depot with Joel H. Litch, a sash and blind manufacturer. In 1846 and 1847 Fiske appears on the Worcester Tax Records and is taxed on $300 worth of machinery and stock in trade. Another indication of his progress was his marriage on December 9, 1846 to Clara M. Wood.

Worcester, like so many other New England manufacturing towns had an association of mechanics which occasionally sponsored exhibits of products of all sorts. The First Exhibition of the Worcester County Mechanic's Association occurred in September of 1848. Although Isaac Fiske did not enter instruments, he was one of six people on the judging committee for musical instruments.[35] In the second exhibition held in 1849 three cornets by Fiske won a silver medal. Following is the judge's report:

> 1609 (a.b.c.) Cornets – (B-flat.) Isaac Fiske, Worcester. Very much admired for beauty of make and finish, and the light, elastic, and ready action of the valves. The tone is excellent. There is an improvement in the working of the valves, by which the execution of rapid passages and trills may be accomplished with great ease and perfection. Silver Medal[36]

Throughout his working life, improving the action of the valves and smoothing the air flow through the instrument were to be Fiske's major goals. The improvement described here was probably Fiske's version of the Vienna twin-piston valve produced by several American makers during the 1840s. A valve bugle in e-flat by Fiske dating from about 1848 is now in the collection of Mr. Fred Benkovic, Wauwatosa, Wisconsin (figure 43). It is made of German silver in upright shape and has three Vienna twin-piston valves manipulated by finger rods with coil springs. The arrangement of rods and enclosed coil springs might have been the improvement referred to since all other similar American valves of this type use a simple lever and flat spring arrangement. The valves are arranged with the shortest valve first, an arrangement still fairly common during the 1840s.

From 1851 until 1853 Fiske's shop was located in a building at 236 Main shared by Alvan and Albert Allen, piano teachers and dealers, and Milton M. Morse, maker of Seraphines and Aeolians (reed organs).

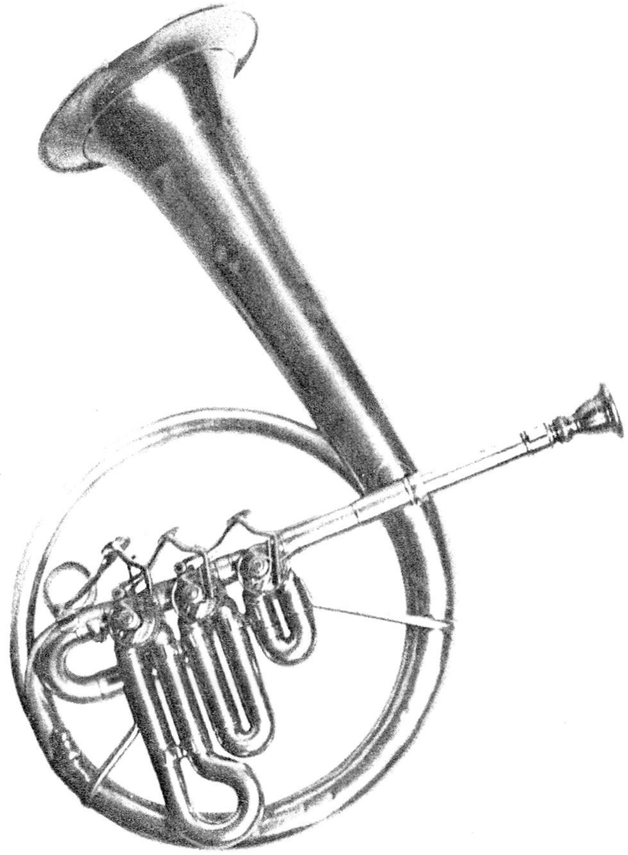

Fig. 44 – Bugle in e-flat, circular, by Isaac Fiske, circa 1850. German silver. Three rotary valves, half-step valve first. (Photo courtesy of Fred Benkovic, Wauwatosa, Wisconsin)

The next Exhibition of the Worcester County Mechanic's Association held in 1851 included three more instruments by Fiske, one of them an alto horn.

> No. 1508. A case of Brass Instruments, by Isaac Fiske, Worcester, containing 1 F cornet, 4 valves; 1 B-flat cornet, 3 valves; 1 E-flat alto horn, 4 valves.
>
> These beautiful instruments are of the highest order. By a peculiar construction of the valves, which have a rotary motion, the manufacturer has succeeded in producing a very short and light action, more than ordinarily rapid and elastic. The instruments have a superior tone, and are remarkable for a liquid blending of the notes, so desirable in Legato passages where all brass instruments are most defective. The committee feel great satis-

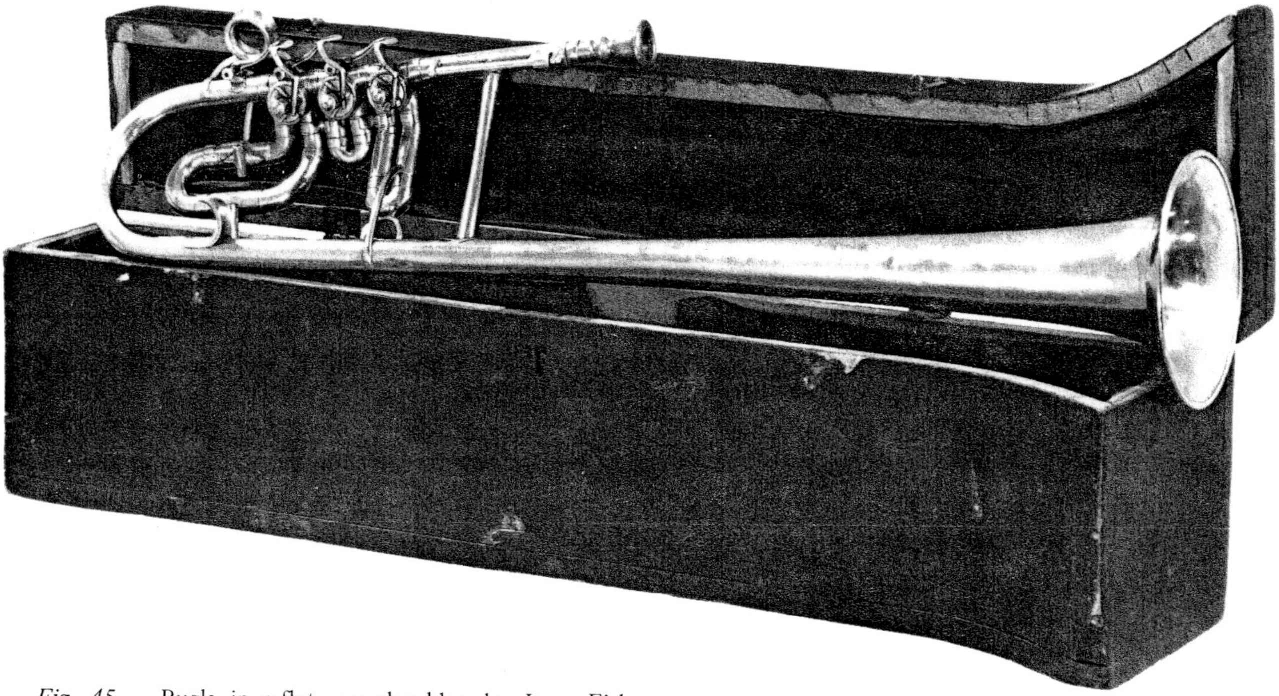

Fig. 45 – Bugle in e-flat, overshoulder, by Isaac Fiske, circa 1852. Brass. Three rotary valves, with case. (Photo courtesy of the Henry Ford Museum, Dearborn, Michigan)

Fig. 46 – Bugle in e-flat, by Isaac Fiske, circa 1855. German silver. Three rotary valves. (Photo courtesy of the Henry Ford Museum, Dearborn, Michigan)

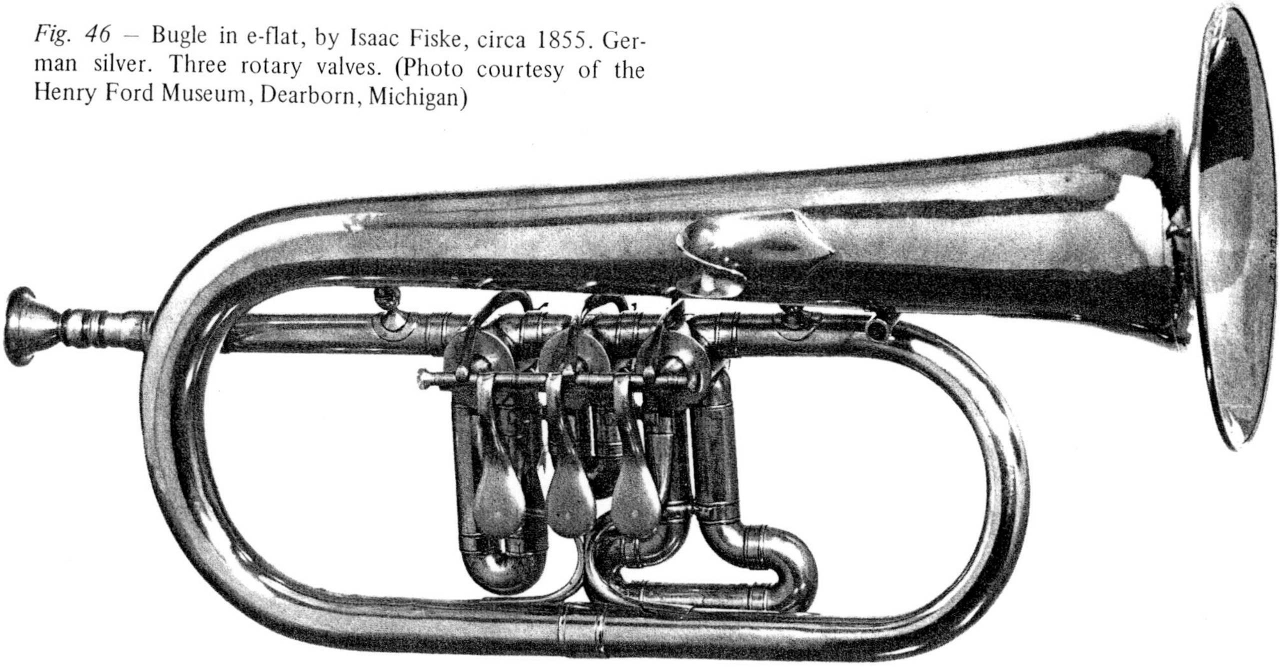

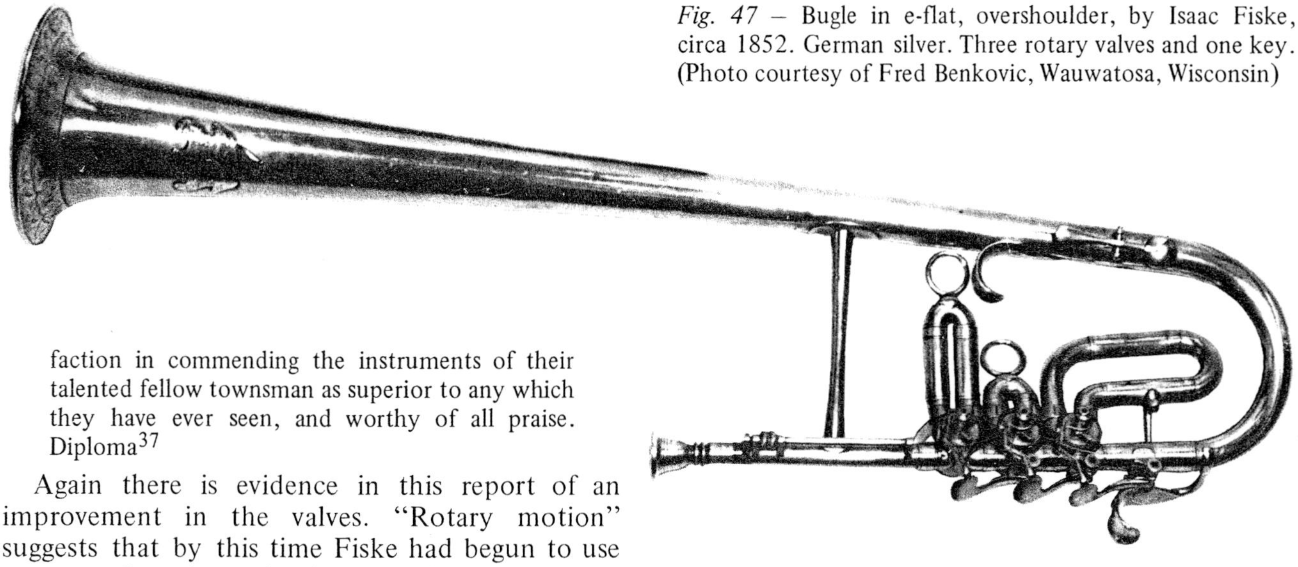

Fig. 47 – Bugle in e-flat, overshoulder, by Isaac Fiske, circa 1852. German silver. Three rotary valves and one key. (Photo courtesy of Fred Benkovic, Wauwatosa, Wisconsin)

faction in commending the instruments of their talented fellow townsman as superior to any which they have ever seen, and worthy of all praise. Diploma[37]

Again there is evidence in this report of an improvement in the valves. "Rotary motion" suggests that by this time Fiske had begun to use rotary valves instead of the earlier twin-piston variety. The improvement described could well be the introduction of string linkage recently developed by Thomas D. Paine of Woonsocket, Rhode Island and soon to be standard on all American rotary valve instruments. A circular valve bugle in e-flat by Fiske also in the collection of Mr. Benkovic shows this type of valve arrangement (figure 44). It is made of German silver and has three rotary valves turned by string linkage. Again the shortest valve is first. Two other valve bugles in the more common overshoulder and trumpet shapes with the usual order of valves are shown in figures 45 and 46.

Other makers, early in the 1850s, were experimenting with instruments having both valves and keys. Both Thomas D. Paine and E. G. Wright are known to have made examples. Fiske also made what must have been a half-hearted attempt at designing an instrument of this type. It is an overshoulder valve bugle with one key, now in the collection of Fred Benkovic (figure 47). The key was probably intended to aid in the production of high *c'''* and is in the same approximate position as key 12 on a 12-key bugle. In this instance it does not provide any advantage and, as far as is known, Fiske did not continue making them.

By this time Fiske's reputation as a fine maker of brass instruments had spread beyond the immediate locality. *Dodworth's Brass Band School* published in 1853 by Allen Dodworth, a well known band leader of New York City, included an illustration of an Isaac Fiske circular cornet with the following note: "the engraving is from one made by Mr. Isaac Fiske of Worcester, Mass., a maker of deserved celebrity; as few can be found anywhere of more perfect workmanship."[38] An instrument similar to the Dodworth illustration is shown in figure 48.

Growth in population and industry in Worcester during the 1840s and 1850s was spectacular. From a population of 7,497 in 1840, Worcester reached 17,049 in 1850 and 24,960 in 1860. The increase of industry is said to have stemmed in part from the availability of rental manufacturing space that included power. Beginning as early as 1832 with the old Court Mill, this kind of space was available.

> The means thus afforded to individuals with limited capital to begin manufacturing unencumbered with an expensive plant making it possible for a small business to be conducted with profit, is one of the chief causes of the diversity of industries which makes Worcester uniformly prosperous and creates a thrifty and permanent class of working people.[39]

One of the largest renters of manufacturing space in Worcester was W. T. Merrifield, whose buildings, powered by a large steam engine, housed twenty to thirty firms in the early 1850s. In 1854 Isaac Fiske moved from the music store on Main Street to a shop in the Merrifield buildings. His move proved to be very ill-timed, however, for a few months later the entire building complex burned to the ground. Fiske lost a reported $1,000 worth of stock and tools, only $500 of which was covered by insurance.[40]

Fiske's business must have prospered the last few years before the fire for, far from being destitute after his loss, he appears wealthier than ever. In 1855 he bought two lots on Piedmont Street, one of which would be his home for the rest of his

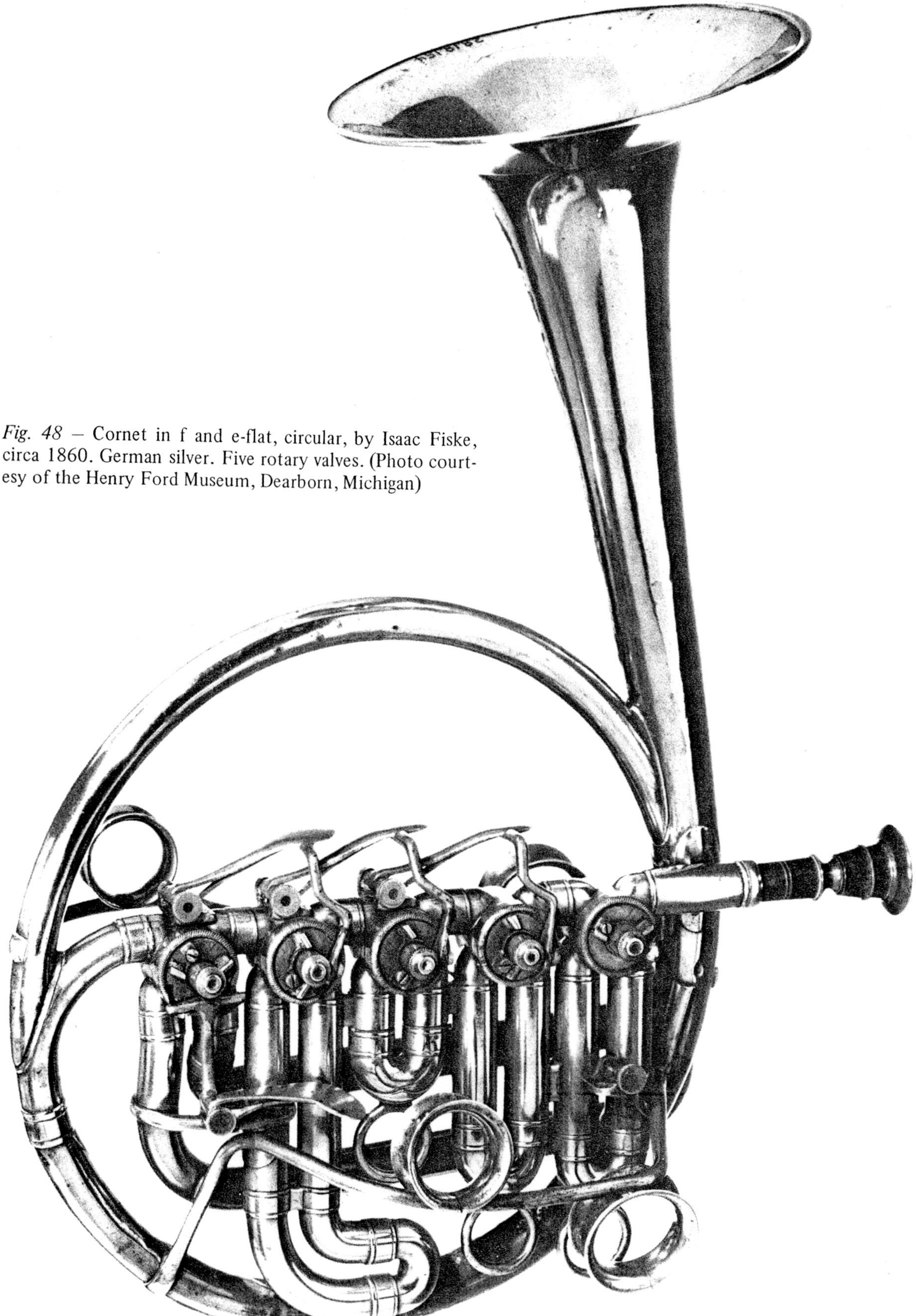

Fig. 48 – Cornet in f and e-flat, circular, by Isaac Fiske, circa 1860. German silver. Five rotary valves. (Photo courtesy of the Henry Ford Museum, Dearborn, Michigan)

FISKE'S

CORNET BAND,

M. ARBUCKLE, LEADER AND DIRECTOR,

—WILL GIVE A—

GRAND CONCERT,

Assisted by

MISS JENNIE TWICHELL, VOCALIST,

And

HERR SCHREIBER,

The greatest Cornet Player now living,

AT MECHANICS HALL,

On WEDNESDAY Evening, September 8th, 1858.

PROGRAMME:

Part First.

1. SCENA ED ARIA—"D'Amor Sull ali Rose," from Ill Trovatore, ... *Verdi.*
CORNET BAND.
2. TERZETTO—from Norma, ... *Bellini.*
SERENADE BAND.
3. CAVATINA—"Ah si Pera," from Donna dell Lago, *Rossini.*
Miss Jennie Twichell.
4. SOLO, CORNET-A-PISTON—Grand Concerto, "Souvenir de Leamington," . . . . . . *Schreiber.*
Herr Schreiber.
5. KATY DID POLKA, ... *Jullien.*
ORCHESTRA.
6. VARSOVIANA, ... *Stasny.*
ORCHESTRA.

Part Second.

1. LA BELLE SUISSE WALTZ, ... *D'Albert.*
CORNET BAND.
2. POTPOURI—from Ill Trovatore, Introducing Anvil Chorus &c. . . . . *Verdi.*
CORNET BAND.
3. BALLAD—Kitty Tyrrell, ... *Glover.*
Miss Jennie Twichell.
4. SOLO, CORNET-A-PISTON—Fantasia on Katy Darling, *Schreiber.*
Herr Schreiber.
5. VALSE L'HIVER—from "Les Vepres Sicilliennies," *Laurent.*
ORCHESTRA.
6. GALLOP DE BRAVURA, ... *Schulhoff.*
ORCHESTRA.

TICKETS TWENTY-FIVE CTS.

To be had at Marsh's Music Store, and at the door of the Hall.

Doors Open at 7 1-4; to Commence at 8 o'clock.

Printed by E. R. Fiske, Foster Street, Worcester.

Fig. 49 – Handbill advertising a concert of Fiske's band, September 8, 1858. (Photo courtesy of the Worcester Historical Society)

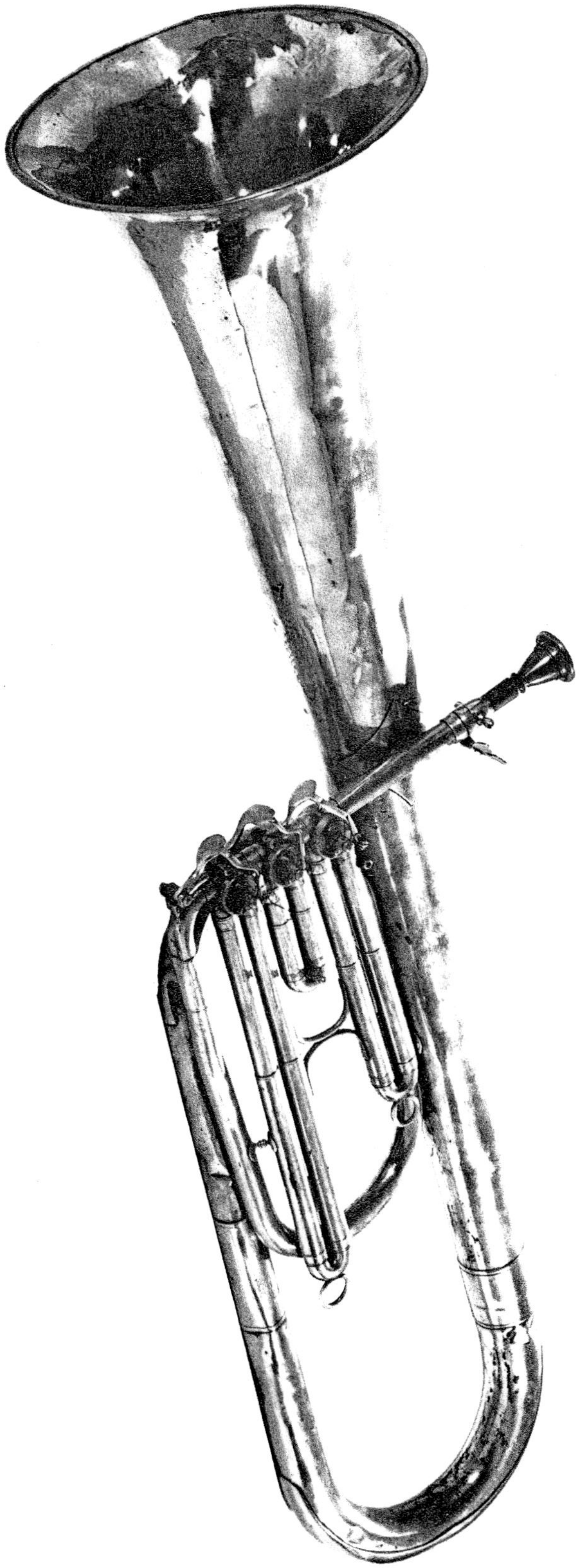

Fig. 50 – Baritone in B-flat, by Isaac Fiske, circa 1855. German silver. Three rotary valves. (Photo courtesy of the Henry Ford Museum, Dearborn, Michigan)

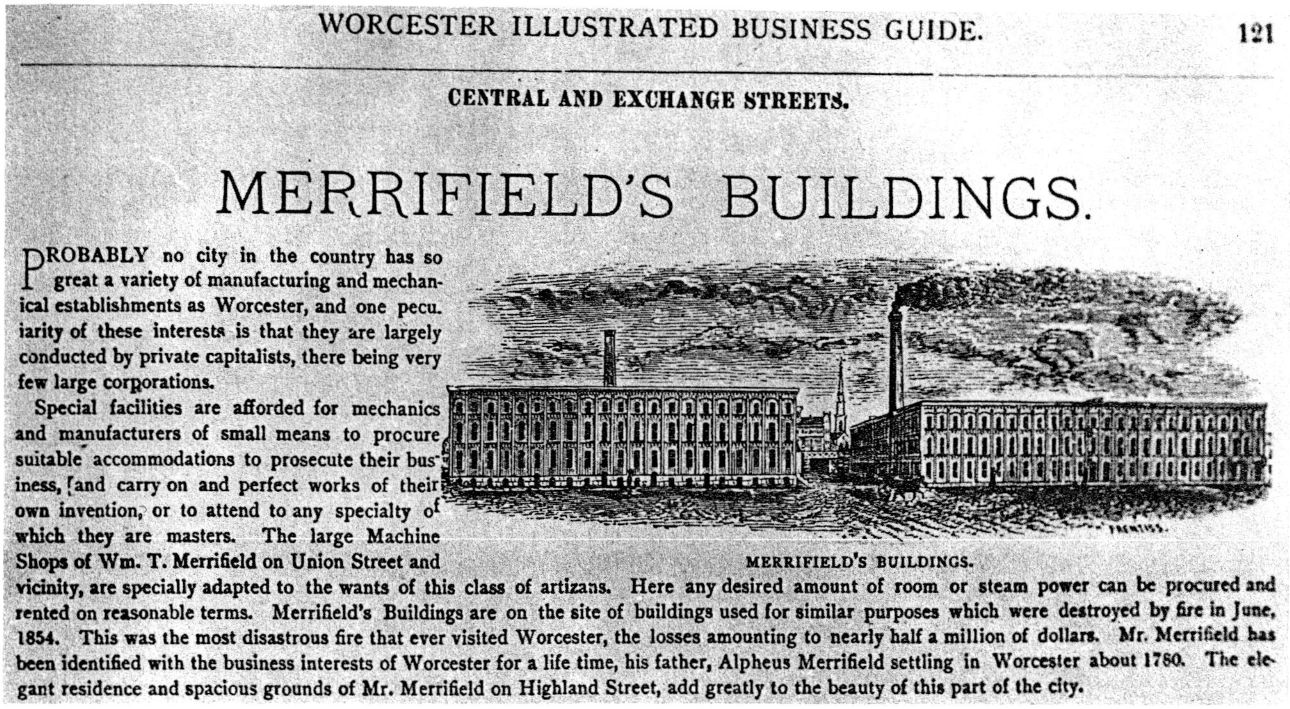

WORCESTER ILLUSTRATED BUSINESS GUIDE. 121

CENTRAL AND EXCHANGE STREETS.

MERRIFIELD'S BUILDINGS.

PROBABLY no city in the country has so great a variety of manufacturing and mechanical establishments as Worcester, and one peculiarity of these interests is that they are largely conducted by private capitalists, there being very few large corporations.

Special facilities are afforded for mechanics and manufacturers of small means to procure suitable accommodations to prosecute their business, and carry on and perfect works of their own invention, or to attend to any specialty of which they are masters. The large Machine Shops of Wm. T. Merrifield on Union Street and vicinity, are specially adapted to the wants of this class of artizans. Here any desired amount of room or steam power can be procured and rented on reasonable terms. Merrifield's Buildings are on the site of buildings used for similar purposes which were destroyed by fire in June, 1854. This was the most disastrous fire that ever visited Worcester, the losses amounting to nearly half a million of dollars. Mr. Merrifield has been identified with the business interests of Worcester for a life time, his father, Alpheus Merrifield settling in Worcester about 1780. The elegant residence and spacious grounds of Mr. Merrifield on Highland Street, add greatly to the beauty of this part of the city.

MERRIFIELD'S BUILDINGS.

Fig. 51 – The Merrifield Buildings in Worcester, Massachusetts. (Photo courtesy of the Worcester Historical Society)

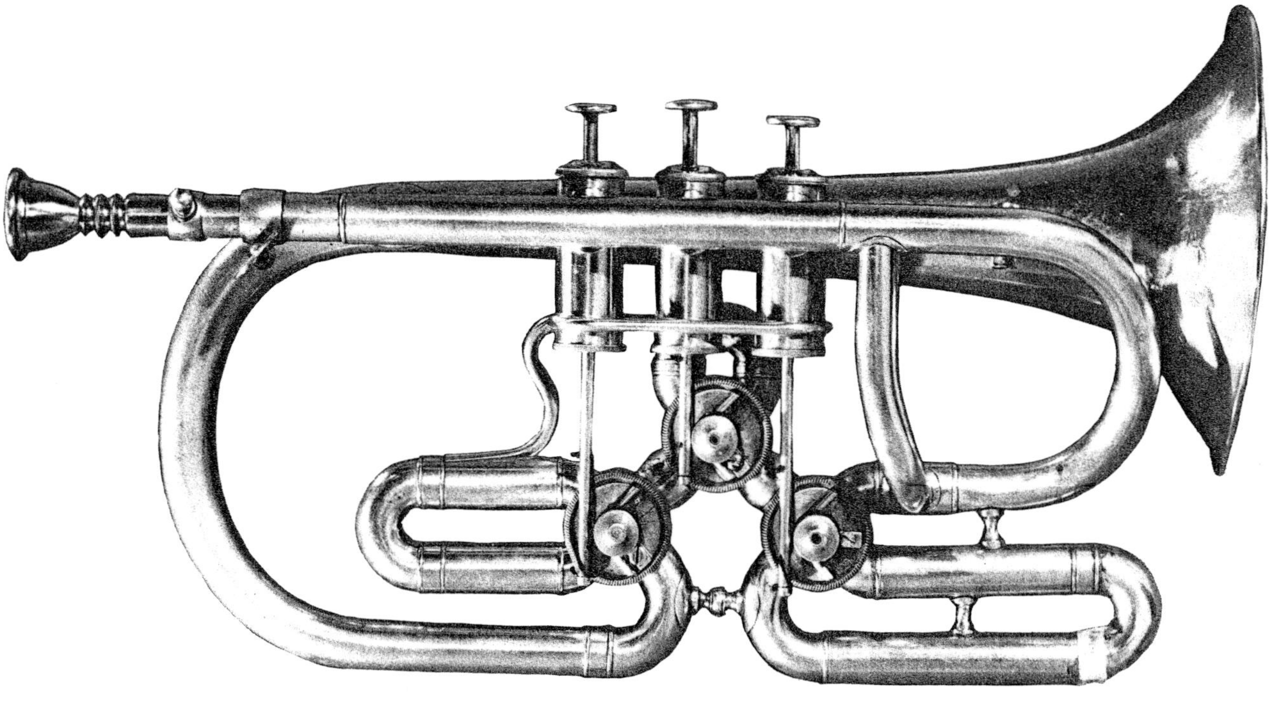

Fig. 52 – Cornet in e-flat, by Isaac Fiske, circa 1880. German silver. Fiske's patented valve arrangement. (Photo courtesy of the Henry Ford Museum, Dearborn, Michigan)

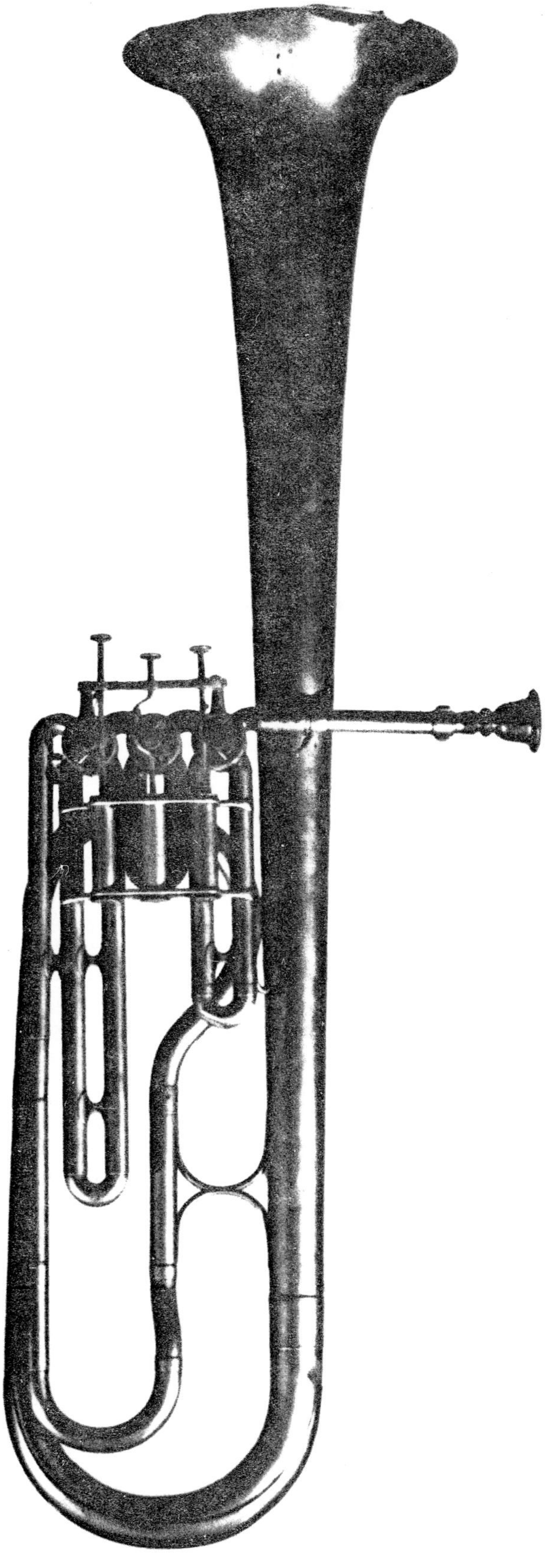

Fig. 53 – Baritone in B-flat, by Isaac Fiske, circa 1870. Brass. Fiske's patented valve arrangement. (Photo courtesy of Dr. Robert M. Rosenbaum, Scarsdale, New York)

life, and in 1857 he purchased another piece of property nearby on Hammond St. The year 1857 also saw the formation of Fiske's Cornet Band and the arrival of Matthew Arbuckle. A handbill announcing one of their concerts is shown in figure 49. In the Fourth Exhibition of the Worcester County Mechanic's Association that year, the band entered a complete set of 13 Fiske instruments including every size from E-flat bass to e-flat soprano. A baritone horn of about this date is shown in figure 50.

> 1382. Set of German Silver instruments. Consisting of 13 pieces. Manufactured by Isaac Fiske of Worcester. Exhibited by Fiske's Cornet Band, Worcester. Their perfect and easy working valves, and their excellent quality of tone, toegether with their external beauty of proportion (setting aside the testimony of a world of witnesses to the fact that Mr. Fiske has long since forgotten how to make a second rate instrument) conspire to render them worthy of a Silver Medal.[41]

In 1856 Fiske paid taxes on $1,000 in ma-

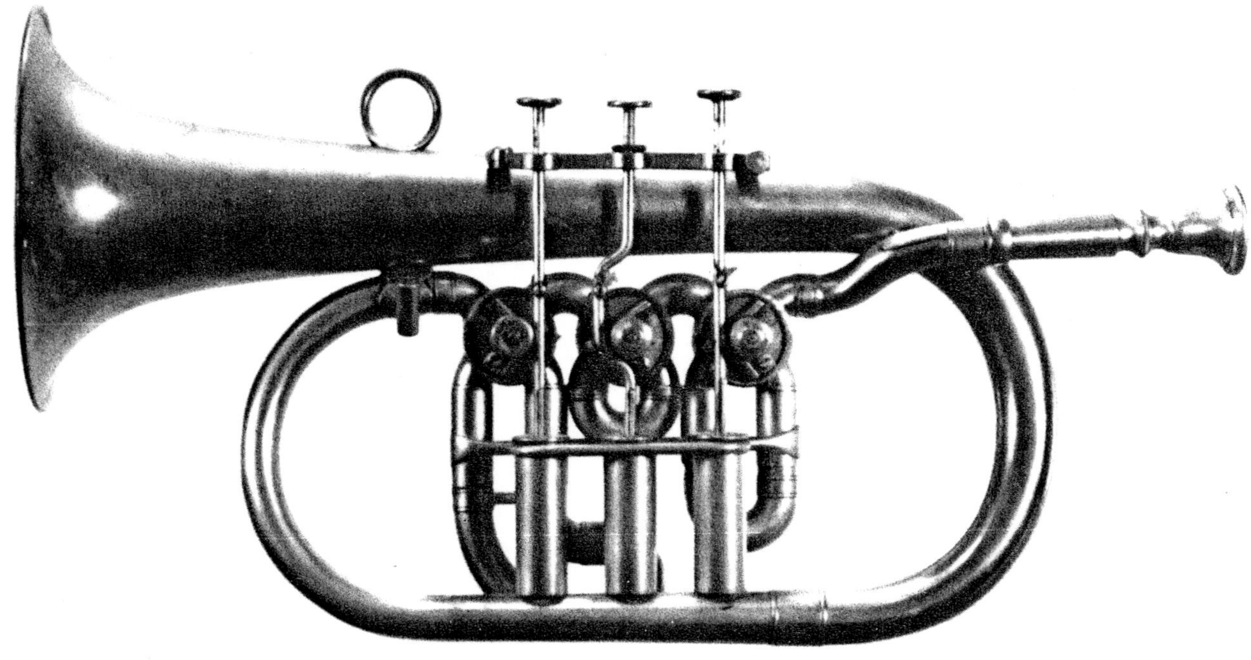

Fig. 54 – Cornet in e-flat, by Isaac Fiske, circa 1870. German silver. Fiske's patented valve arrangement. (Photo courtesy of Alfred F. Wood, Westerly, Rhode Island)

chinery and stock in trade and $2,100 in real estate. The following year these values had risen to $1,200 and $2,800 respectively.

By 1859 the Merrifield buildings had been rebuilt and enlarged to three stories high. Over a mile of shafting carried power from a 350-horse-power "Lawrence" steam engine to every room and shop. Isaac Fiske set up shop in the new building with six employees. Although he moved occasionally to different areas of the building, Fiske remained a tenant of the Merrifield Co. until his retirement in 1887. (Figure 51)

When Matthew Arbuckle left in 1860, the Fiske Cornet Band was disbanded and the set of Fiske instruments sold to a board of trustees who formed a new band called the National Band. The details of the arrangement were spelled out in great legal detail and properly recorded with the registry of deeds book 635, page 2, 3. The war years were prosperous ones and the value of Fiske's taxable property rose to $3,650 in business assets and $2,500 in real estate by 1865.

On October 30, 1866 Isaac Fiske obtained his first United States patent, No. 59204. In it he sought to protect several improvements on brass instruments. The most interesting part of this patent deals with an arrangement of cornet valves and fingering mechanisms. Fiske placed three rotary valves along the bottom of the instrument turned with string linkage by three vertical rods. Each of the rods passed up through a cylinder containing the return spring, to the finger button in the usual position (figure 52). Another unusual arrangement based on the same idea but with spring casings below the valves is illustrated in figures 53 and 54.

The advantages claimed were shorter, more direct movement and straight up-and-down motion instead of the arcing motion of the usual rotary valve lever. The idea was not entirely new, as Joseph Higham of Manchester, England, had patented a similar arrangement in 1857, British Patent 123. A cornet of c. 1850 by Robinson & Bussell, Dublin, in the collection of Henry Meredith (London, Ontario, Canada) also has similar valves. Fiske's use of string linkage was perhaps a little better than the mechanical arrangement used by Higham (figure 55). It allowed the finger rod to travel a shorter distance and produced less side pressure on the bottom of the rod. Although Fiske stuck with his idea for many years, the slight advantage was not worth the weight and cost of the extra equipment it required, and the arrangement was eventually discarded.

Another part of this patent was the idea of "interposing rubber or some other suitable elastic substance between the attachments . . . of the main pipe with the bell . . . of a wind instrument to give greater freedom to the vibrations of the bell . . ."

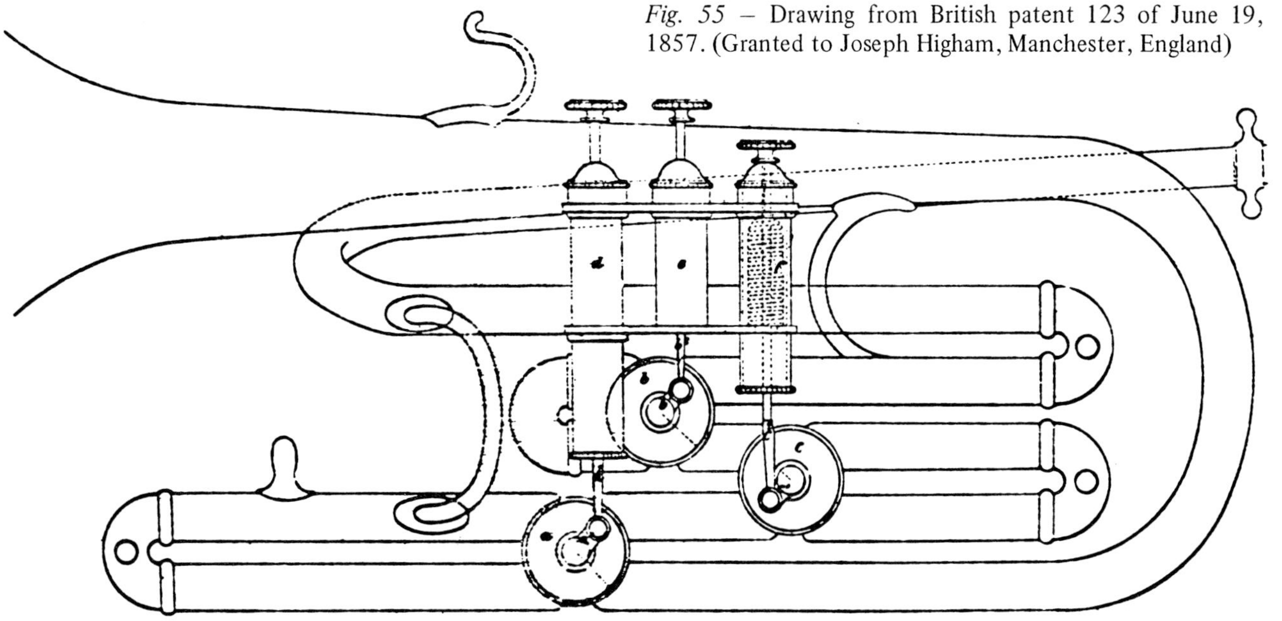

Fig. 55 – Drawing from British patent 123 of June 19, 1857. (Granted to Joseph Higham, Manchester, England)

Other American makers were also concerned about this aspect of brass instrument construction during the 1860s. Louis Schreiber of New York patented a cornet design in 1865 (No. 49925) with the tubing arranged so that the player's hand would not touch the bell and damp its vibrations. B. F. Richardson of Boston designed a cornet with the bell hanging under the valves without braces or other attachments for the same reason.[42] None of these ideas was found to be worth continuing.

Fiske's next patent was No. 70824 of 1867 and covered a manufacturing process by which small curved tubes or crooks could be formed easily from one piece of sheet metal. A flat piece of metal was cut in a particular shape, formed on a die and joined on the inside of the curve and part way from each end on the outside of the curve. None of the instruments found so far appear to have crooks formed by this process, so it can only be assumed that it was not very successful. Curved tubes in brass and copper were usually formed by bending, and in German silver, which is harder to work, by shaping two halves in a die and soldering them together. Later instruments were made of brass and then plated covering all evidence of seams and other marks of construction.

The Fifth Exhibition of the Worcester County Mechanic's Association was held in September of 1866 and Fiske entered an improved cornet described in the following report of the judges:

> A soprano cornet made and contributed by Isaac Fiske of Worcester. Mr. Fiske in this instrument introduces what the judges believe to be one of the best improvements ever introduced to any brass instrument. It is simply a newly shaped valve by which all sharp angles in the sounding tube are avoided, thus rendering the valve tones, nearly or quite as pure as the open or natural tones . . . highest diploma . . .

This improvement was subsequently protected by the United States patent 74,331 of 1868.

> The invention consists in the construction and arrangement of the valve-passages with reference to their connections with the pipe, by which, through the valve passages and pipes, both in the open and valve-tones, a continuous wind-passage of uniform diameter, free at all times from sharp turns and corners, is formed.
>
> Also in the construction and arrangement of a valve and valve-case, by which the instrument may be changed from one to another key in such manner that the same quality of tone may be produced in either key, and without increasing the number of crooks or turns in the wind passage.

Fiske was about ten to fifteen years behind similar work done in Europe. Gustave Besson patented a set of piston valves in 1855 in which the main accomplishment was the maintaining of the same dimensions of bore through all windways and in all possible combinations of the pistons. Dr. J. P. Oates of Lichfield, England designed what he called equitrilateral valves very similar to Fiske's change of key valve in 1851.[43] Nevertheless American makers were stimulated to better design by the Fiske experiments.

The 1860 United States census of manufacturing

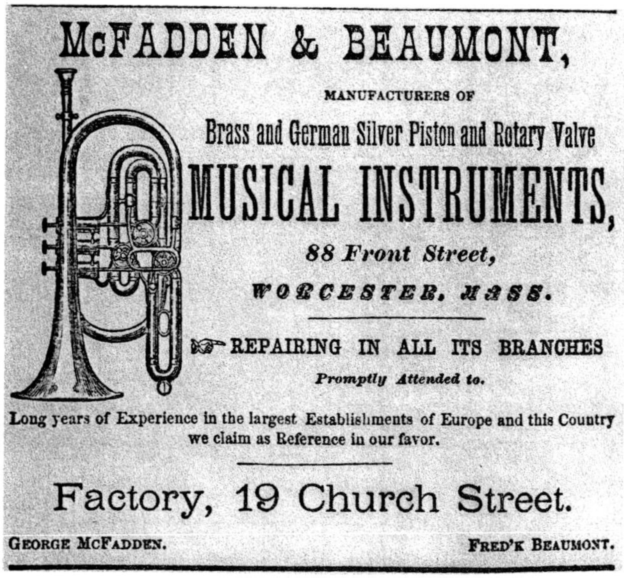

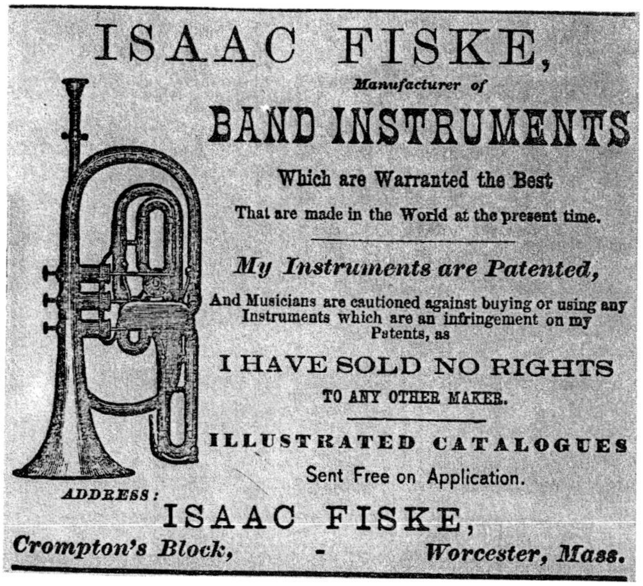

Fig. 56 – Advertisements from the 1873 Worcester City Directory during the dispute between Isaac Fiske and McFadden and Beaumont.

2207

ILLUSTRATED CATALOGUE

— OF —

Musical Instruments,

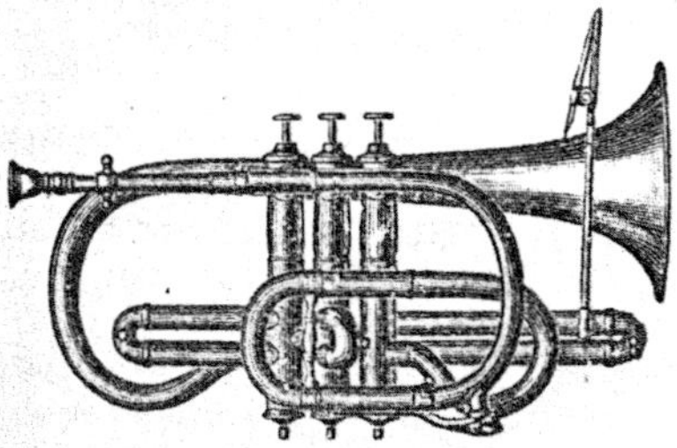

MANUFACTURED BY

ISAAC FISKE,

Crompton's Block, Mechanic St.,

WORCESTER, - - MASS.

July, 1881.

E. R. FISKE & SON, PRS., WORCESTER.
1881.

INTRODUCTION.

Those who have not seen or had a chance to try my Instruments of late, are informed that I have, in the last six months, made *an improvement in my models*, so that now I am *willing* and anxious to send them for trial to any one who wants to buy a *first-class Instrument.* I only require a deposit with the Express Company, of the amount of bill and charges, and after three days trial, you can return it to the Express Company and *take your money*; or if it proves to be satisfactory and all that I claim for it, you can keep it and send me the money.

A novice cannot judge of what he does not know, therefore if you do not play yourself, get some professor or other person who does play the instrument, and is *honest* and *capable of judging*, to try it for you, and write me just what he says.

I do not compete in price with a large proportion of Instruments imported or made in this country, but I only ask you to *compare mine* with the *very choice* ones, then compare the prices. I do not put mine by the side of those that are imported for from six to twelve dollars each, and sold here for from fifteen to thirty dollars, and pronounced superior. I defy competition and guarantee my Instruments perfect in every particular.

Fig. 57 – Illustrated Catalogue of Instruments manufactured by Isaac Fiske. (Photos courtesy of the Worcester Historical Society)

4

SILVER PLATING.

These prices are for Plating only, and will be added to price of Instrument in Brass.

E♭ Cornet, triple silver plate, satin finish,							$12 00
B♭	"	"	"	"	"	"	12 00
E♭ Alto,*	"	"	"	"	"	"	15 00
B♭ Tenor,	"	"	"	"	"	"	20 00
B♭ Baritone,	"	"	"	"	"	"	25 00
E♭ Bass,	"	"	"	"	"	"	35 00

These prices are for *first-class* Plating only.

Engraving and Gilding can be put on at any price from five to fifty dollars.

IMPROVED MOUTH PIECES.

	Brass.	G. Silver.	Plate.
E♭ Cornet, - -	$1 25	$1 50	$1 75
B♭ " - - -	1 25	1 50	1 75
E♭ Alto, - -	1 50	1 75	2 00
B♭ Tenor and Baritone,	1 50	1 75	2 00
B♭ Bass, - - -	1 75	2 00	2 25
E♭ " - -	2 00	2 50	2 75

5

E♭ CORNET.

Side Action, Patent Rotary Valves. Intended for leaders of brass bands who prefer the side action.

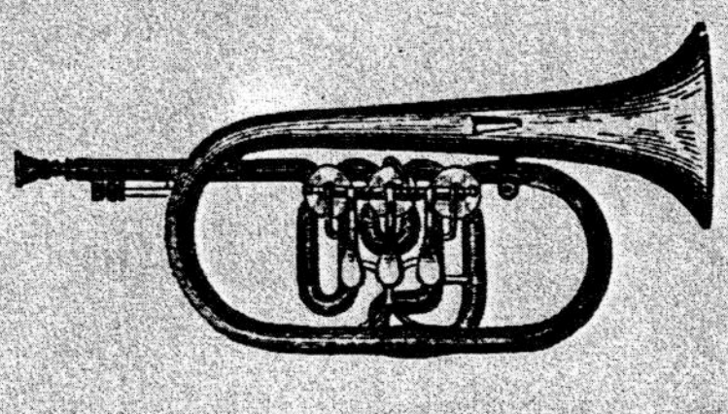

This Instrument stands above all others of this class.

Price for Brass, - - - - -	$45 00
" " " Silver Plated, - -	57 00
Gold inside of bell, extra, - - -	5 00
Black Walnut Box, Velvet inside, -	10 00

6

E♭ CORNET.

Piston Valve. This Instrument stands at the head of *all* E♭ *Cornets. Try it and see.* Leaders of bands are respectfully requested to give this Instrument a *trial.* Water Key and Music Holder.

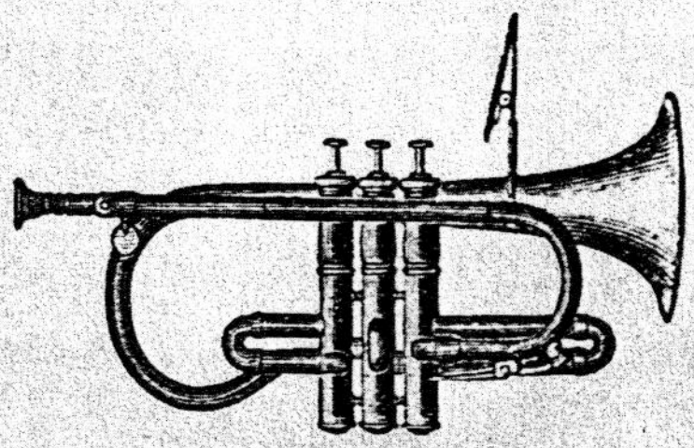

Price for Brass, - - - - -	$45 00
" " " Silver Plated, - -	57 00
Gold inside of bell, extra, - -	5 00
Black Walnut Box, Velvet inside, - -	10 00

7

E♭ CORNET.

Rotary Valve, Piston Action, only three-eighths of an inch long. Water Key and Music Holder.

Price for Brass, - - - - -	$45 00
" " " Silver Plated, - -	57 00
Gold inside of bell, and all parts Gilded, extra,	20 00

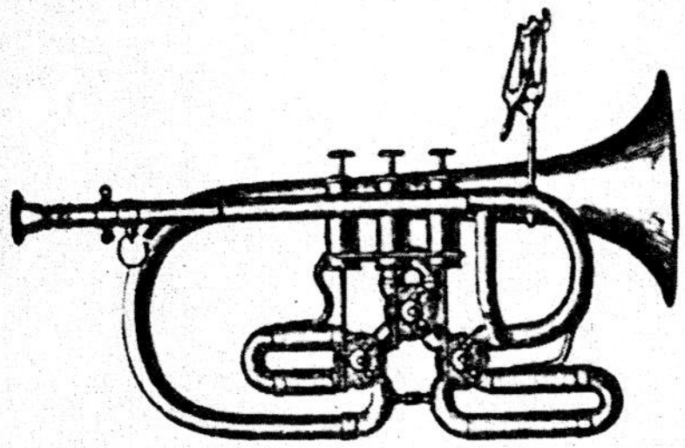

IMPROVED E♭ CORNET.

Rotary Valve, Piston Action, three-eighths of an inch in length. The *improvement* consists in the direction of the wind passage. The wind continues in one direct course, and is *not reversed,* which makes a great difference in the quality of tone, and the ease with which it is produced.

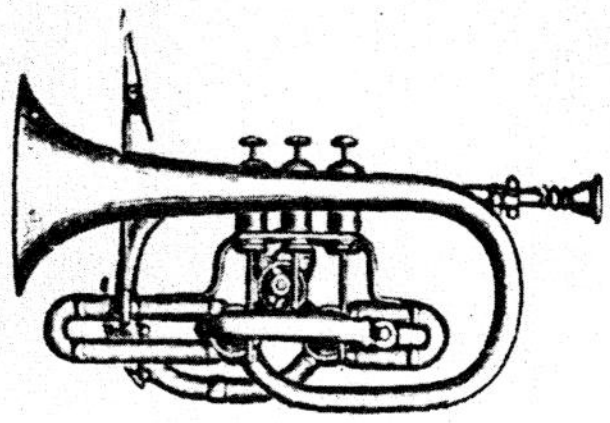

B♭ CORNET.—IMPROVED MODEL.

Is warranted to be *perfect* in every particular. The form of model is such that the *tone of each note is perfect,* and that every note in the scale, whether *open or valve tone,* comes *precisely alike,* and *blows remarkably easy. Teachers* are particularly requested to *try this Instrument.* Double Water Key, and set piece for A, and Music Holder.

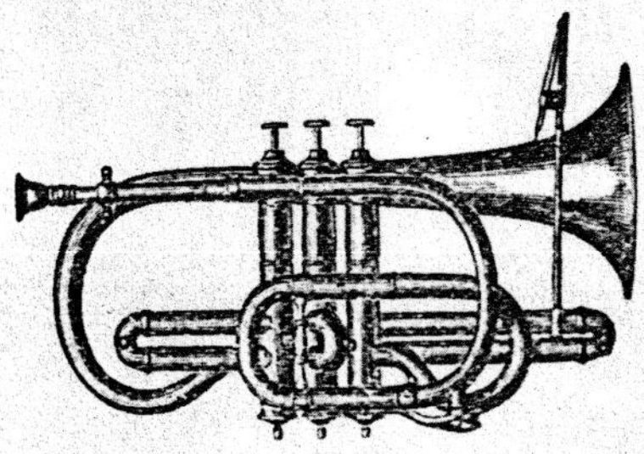

Price for Brass,	- - - -	$55 00
" " " Silver Plated,	- -	67 00
" " " " " and Gilded,		75 00
Black Walnut Box, instrument fitted, lined with Velvet,	- - - -	10 00

B♭ CORNET.

Rotary Valve, Piston Action, only three-eighths of an inch in length, two Water Keys, set piece for A, and Music Holder.

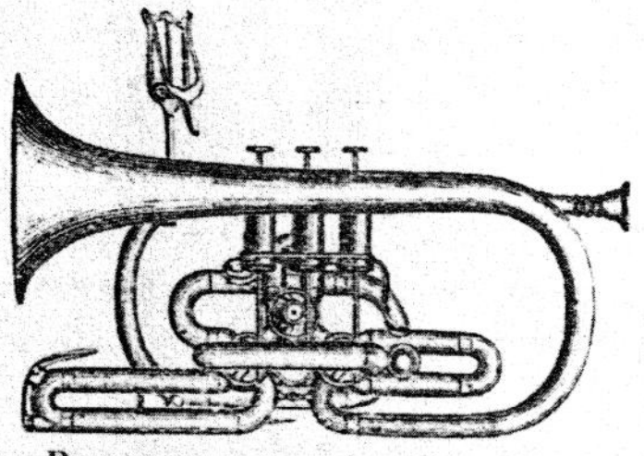

Price for Brass,	- - - - -	$55 00
" " " Silver Plated,	- -	67 00
Inside of bell and points Gilded,	-	75 00

IMPROVED B♭ CORNET.

Rotary Valve, Piston Action, three-eighths of an inch in length. The *improvement* consists in the direction of the wind passage. The wind continues in one direct course, and is *not reversed,* which makes a wonderful difference in the quality of tone and the ease with which it is produced. *Try it.*

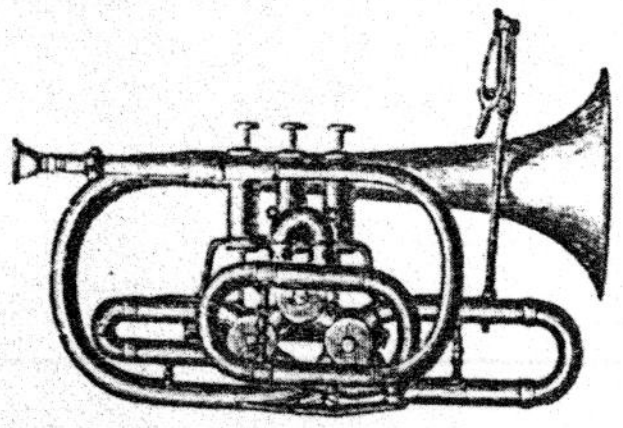

Double Water Key, set piece for A, and Music Holder. Price same as above.

PATENT B♭ TENOR TROMBONE.

Rotary Valve, Piston Action. This Instrument is intended for either Brass Band or orchestra; is the right calibre for *Tenor.*

Price for Brass,	- - - - - - - - - -	$65 00
" " " Silver Plated,	- - - - - - - - -	85 00
" " " Piston Valve,	- - - - - - - - -	65 00
" " " " Silver Plated,	- - - - - - - - -	85 00

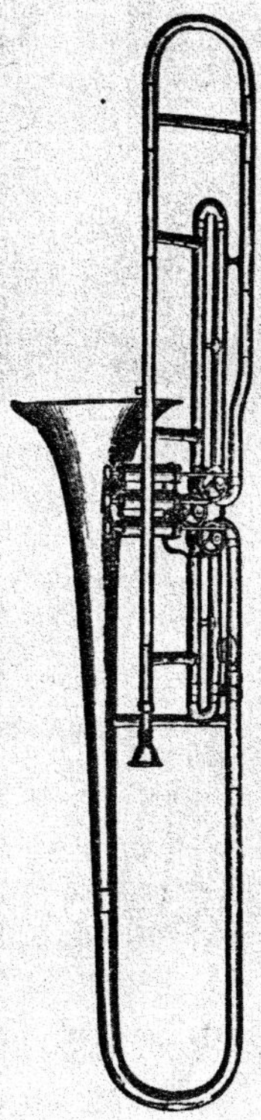

E♭ BASS PISTON VALVE.

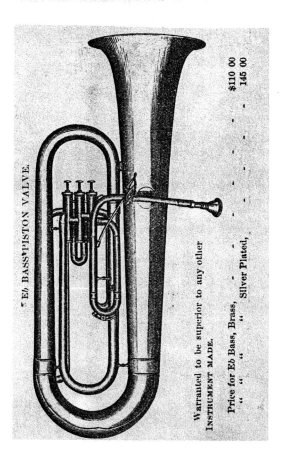

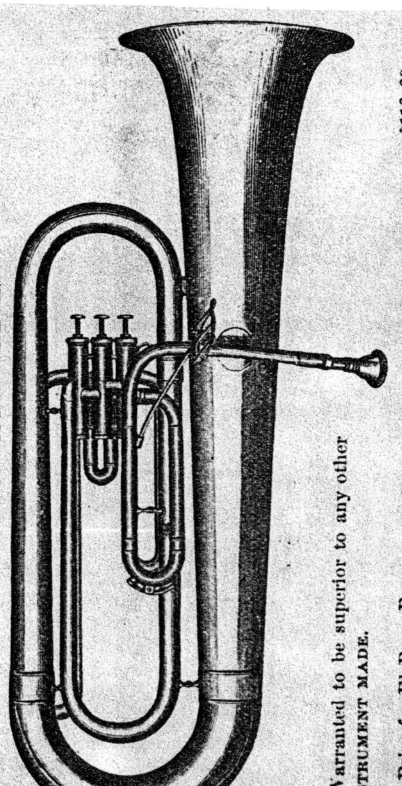

Warranted to be superior to any other INSTRUMENT MADE.

Price for E♭ Bass, Brass,	- - - - - - -	$110 00
" " " Silver Plated,	- - - - - - - -	145 00

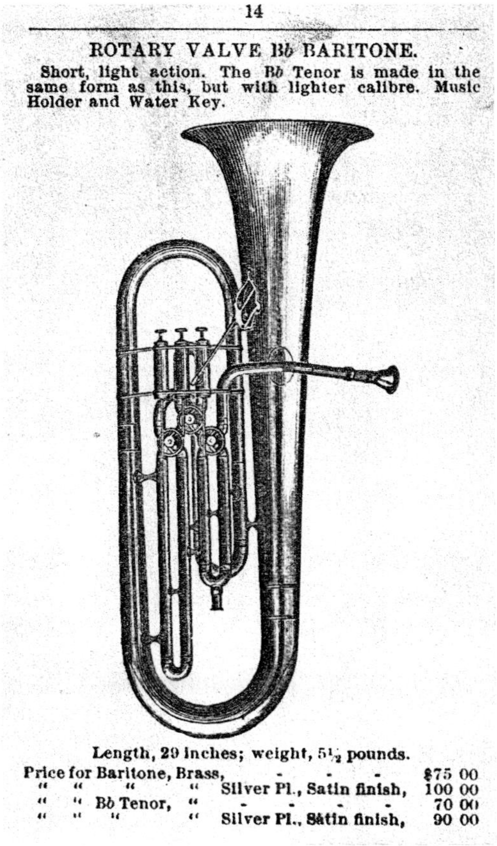

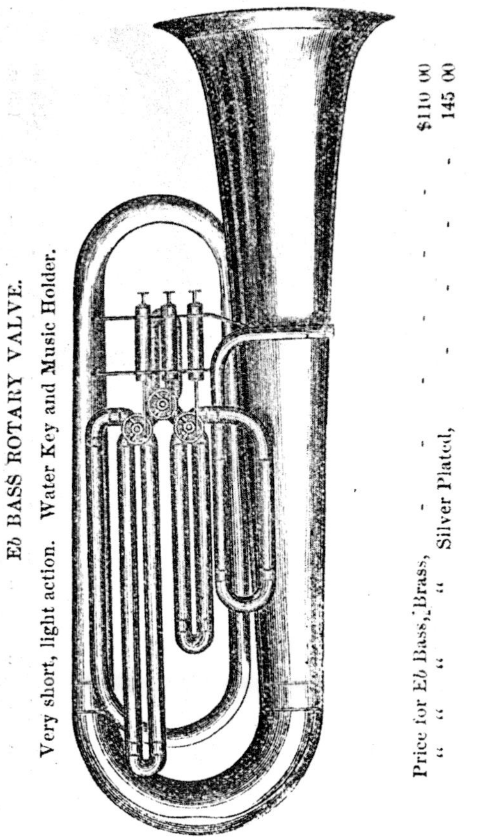

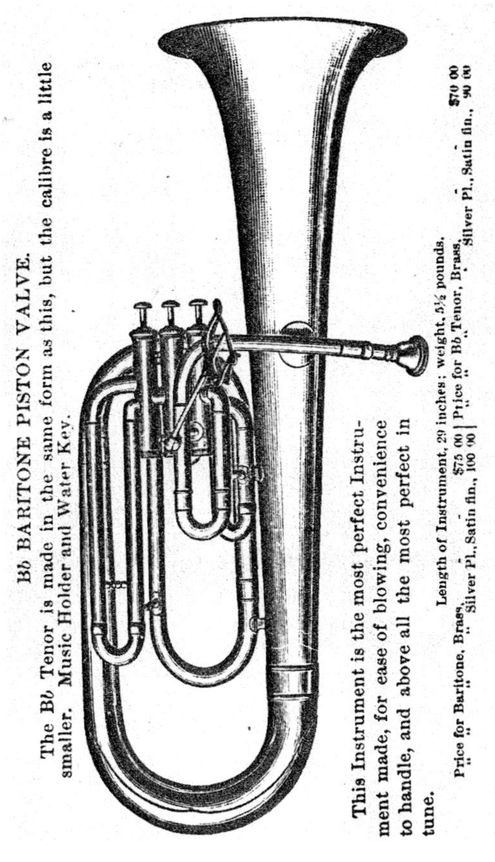

firms reported that Fiske had 10 employees and a payroll of $350 per month. In the early 1870s Fiske's shop employed seven musical instrument makers.[44] Although his work force may have been somewhat larger for a brief time during the Civil War, the firm probably employed no more than a dozen people at any one time.

About 1872 Frederick Beaumont, one of the musical instrument makers employed for several years by Fiske, quit and set up his own shop with George McFadden, a merchant in hosiery and fancy goods. Unfortunately the instruments they made and advertised were identical to those developed and patented by Fiske and a bit of a squabble ensued. In the Worcester City Directory of 1873 advertisements of the two firms appeared on facing pages (figure 56). Although court records reveal no action by Fiske, McFadden and Beaumont were out of business by 1875.

Patents 138,389 and 143,134 were obtained by Fiske in 1873, both covering minor improvements in the layout of the valves to obtain windways with as few sharp bends as possible.

By 1873 Fiske had begun to issue catalogues of his instruments. The earliest one known to exist

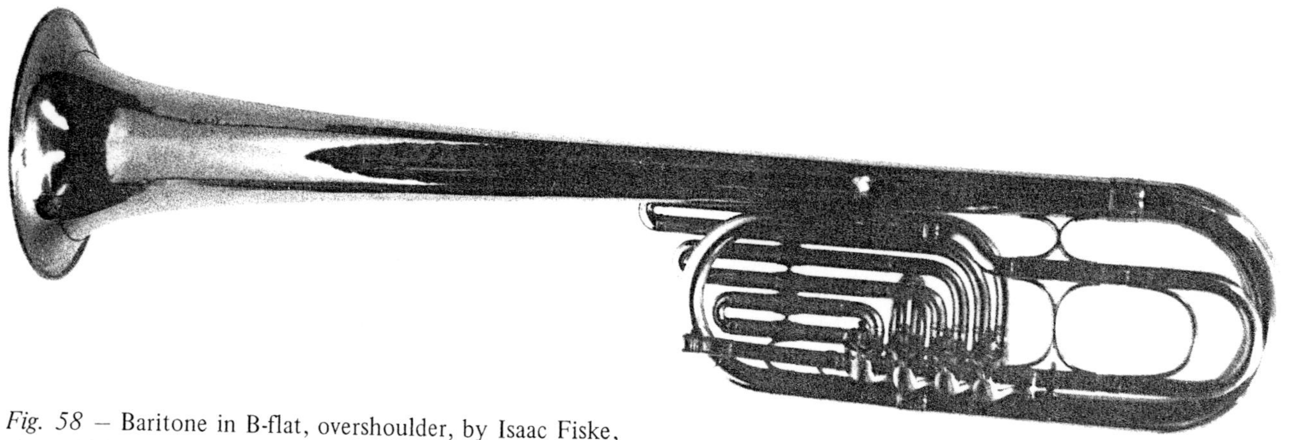

Fig. 58 – Baritone in B-flat, overshoulder, by Isaac Fiske, circa 1860. German silver. Four rotary valves. (Photo courtesy of Don Essig, Central Missouri State University, Warrensburg, Missouri)

is dated 1881 (figure 57). It is interesting that this catalogue does not contain a single example of the overshoulder model made earlier by Fiske and many of his contemporaries. (See figures 45, 47, and 58) A trend toward the Périnet piston valve is also evident in Fiske's offering of either piston or rotary valves on each model. A late Fiske cornet with Périnet piston valves is shown in figure 59.

In addition to the usual information, the 1886 catalogue included a number of testimonials from satisfied customers.

New York, March 6, 1883.

MR. ISAAC FISKE:

Dear Sir.–I have played upon the Instrument which I received from you every day since it arrived, with others, keeping in mind certain qualities: 1st, Resonance; 2d, Tone; 3d, Tune; 4th, Equality.

In regard to resonance I tested it under two heads–carrying power and reverberating–or the quality of its response to the lips.

I have noticed in the Besson Cornets, when in comparison with the Courtois or others, an outsidedness–so to express it–in the effect of the tone or sound coming from it. That is, the ear is impressed as if the sound came from the outside, not muffled by coming from the inside, and distance increases the effect when in comparison with others. This is one of the qualities which gives value to old violins, a quality I am pleased to say belongs to your Instrument, and its responsive power is very delicate, the slightest breath producing sound. These qualities I have never known but one to possess so finely as yours.

2d. The tone is pure, clarion, clear and flexible. Of course you know what I mean by flexible. Some Instruments are stiff, lumpy, the different tones don't seem to flow from one to the other,–they seem to come out jerkey, with effort.

3d. I find it generally in tune. It is not perfect, but he must be a poor blower indeed who cannot play it perfectly in tune.

I am very much pleased with the general easy management of the 3d valve notes which are to be played in tune in all keys without alteration of a slide, a rather strong evidence in its favor, when we bear in mind that G sharp is NOT A-flat.

4th. The general equality between the notes is remarkable. I could play on that Instrument and defy the keenest ear to detect an open from a valve tone. Also, on Instruments generally there is some note or notes which require humoring: that is, they don't blow as they naturally should in sequence to their neighbors. I have not detected that on your Instrument as yet.

Taken all in all, I pronounce it a most excellent Instrument, and it seems to grow in excellence with playing on.

Yours very Respectfully,

H. B. Dodworth.

Harvey Dodworth was a New York band leader and the brother of Allen Dodworth.

The following description of Fiske's business appeared in *Commerce, Manufactures and Resources of the City of Worcester and Environs: A Descriptive Review*, James P. McKinney ed., Worcester: National Publishing Company, 1882, page 75.

Isaac Fiske,

Manufacturer of Band Instruments,
13 Mechanic Street

The United States annually manufactures and imports large quantities of musical instruments and miscellaneous merchandise. The fact that such manufactures and importations grow in number and value from year to year, is pleasing proof of the spread of that musical education which is an evidence of a higher civilization.

The gentleman whose name heads this article,

Fig. 59 – Cornet in B-flat, by Isaac Fiske, circa 1883. Nickel-silver plated brass. Three Perinet piston valves. Photo courtesy of the Henry Ford Museum, Dearborn, Michigan)

er.

may be justly classed among those, who, bringing long practical experience, deep research and study into every detail of their business, attain what can be obtained in no other way–excellence in their chosen business or profession. No concern in this line of business is better or more favorably known, and the workmanship, merits and tone of the instruments here manufactured have gained an enviable reputation through the country.

Mr. Fiske has been engaged in the manufacture of band instruments for about forty years and possesses unusual qualifications for their production.

His premises, located in the Crompton Block, are equipped with all necessary tools, appliances, brazing furnaces, etc., and all the arrangements of the house for the production of the finest instruments in the market are of the most admirable character. He has received testimonials from the highest authority attainable, which declare that for purity of tone and the fine and thorough manner in which they are finished, his instruments are unsurpassed by those of any other manufacturer in the country, or in fact in the world.

Besides manufacturing a full line of Brass and German Silver Band Instruments, Mr. Fiske makes a specialty in making Cornets and instruments for presentation, manufactured either of gold or silver, or both combined, elegantly engraved in original designs, with case complete. He has just delivered a Prize Cornet for the Band Tournament in Maine, which is one of the finest instruments ever turned out, and is valued at $150.

He guarantees any instruments made by him to be perfect in all details of tone, tune and workmanship; and is willing at any time to have them tested for superiority, against any other similar instruments manufactured.

As an authority on band instruments, Mr. Fiske stands very high, and cheerfully renders any information which his extended experience may suggest, whenever requested.

We commend this house to band masters and musicians generally, as one with which to establish relations of the most profitable and pleasant character.

In 1887 Fiske sold his business to C. G. Conn, Inc. and retired. Conn continued operations in Worcester until 1898. Fiske died of apoplexy September 17, 1894.

Isaac Fiske was a first rate craftsman who made excellent instruments throughout a forty-five year career. Quite a number of his instruments survive in various collections attesting to his skill and a fair volume of business. Even though it was about ten to fifteen years behind similar work in Europe, his attempts to improve the action of valves and to design an instrument with clear windways was important to the American brass instrument industry. His business was never as large as Graves & Company of Winchester, New Hampshire, nor as famous as that of E. G. Wright of Boston, but Isaac Fiske made top-notch instruments acclaimed and used successfully by soloists of the day.

The work of Paine, Allen, Wright and Fiske together with that of Samuel Graves helped lay the foundations for an American brass instrument industry. The early period of their activities, from about 1835 to 1855, was devoted to the production of keyed brasses while various valve designs were tried. At this time the first all-brass bands were formed and the key bugle, in the hands of performers like Francis Johnson (1792-1844) and Ned Kendall (1808-1861), was the leading solo voice. The years from about 1855 to 1875 saw a turn from key bugles and ophicleides to valve instruments. Several larger firms were formed and with the earlier makers produced large quantities of rotary valve brasses. Soloists like Matthew Arbuckle (1828-1883) and Patrick Gilmore (1829-1892) continued the American solo brass tradition playing rotary valve bugles and cornets. The first fine professional bands such as Dodworth's and Gilmore's flourished. Graves, Paine, Allen, Wright and Fiske set the standards of inventiveness and workmanship for these developments in America.

Notes

1 Allen Dodworth. *Dodworth's Brass Band School*. New York: H. B. Dodworth, 1853.
2 Thomas Steere. *History of the Town of Smithfield*. Providence: E. L. Freeman & Co., 1881, p. 179.
3 E. Richardson. *The History of Woonsocket*. Woonsocket, Rhode Island: S. S. Foss, 1876, p. 73.
4 Smithfield deed records, Vol. 17, p. 558. Central Falls, R.I., City Center.
5 Ibid., Vol. 18, p. 38.
6 John F. Moore. *The Providence Almanac and Business Directory for the Year 1847*. Providence: J. F. Moore, 1847, p. 107.
7 Concert Program of the American Brass Band of Providence, Rhode Island, March 10, 1851. Rhode Island Historical Society.
8 Dodworth, p. 16.
9 Smithfield deed records, Vol. 32, p. 522.
10 American Brass Band clipping file, Rhode Island Historical Society Library, Providence, Rhode Island.
11 Smithfield deed records, Vol. 37, p. 36.
12 *Leading Manufacturers and Merchants of Rhode Island*. New York: International Publishing Co., 1886, p. 194.
13 North Smithfield, R.I., deed records, Vol. 19, p. 286.
14 Town of Holland, Massachusetts, birth records of 1815, p. 86.
15 Reverend Martin Lovering. *The History of Holland, Massachusetts*. Rutland, Vermont: The Tuttle Co., Marble City Press, 1915, p. 383.
16 Town of Sturbridge, Massachusetts, marriage records, May 7, 1839.
17 *American Journal of Music and Musical Visitor*. Boston, July 16, 1844, p. 343.
18 Norwich, Connecticut, miscellaneous records, Vol. 54, p. 403.
19 *Report of the 15th Exhibition of the Ohio Mechanic's Institute, Held in Cincinnati from September 10 to October 8, 1857*. Cincinnati: by the Institute, 1858, p. 109, 110.
20 Suffolk County, registry of deeds, Vol. 715, p. 234½-236.
21 Op. cit. (Note 15)
22 Ashby, Massachusetts, birth records of 1811.
23 Massachusetts, marriage records for 1864, Vol. 172, p. 116.
24 *The Third Exhibition of the Massachusetts Charitable Mechanic Association at Quincy Hall in the City of Boston, September 20, 1841*. Boston: T. R. Marvin, 1841, p. 84.
25 From the collection of the Tannahill Research Library of the Henry Ford Museum, Dearborn, Michigan.
26 Ibid.
27 "The Boston Brass Band." *Gleason's Pictorial Drawing Room Companion*. Boston, August 9, 1851, p. 225.
28 *Fifth Exhibition of the Massachusetts Charitable Mechanic Association at Faneuil and Quincy Halls in the City of Boston, September, 1847*. Boston: Dutton & Wentworth, 1848, p. 40.
29 Dodworth, p. 16.
30 *Tenth Exhibition of the Massachusetts Charitable Mechanic Association at Faneuil and Quincy Halls in the City of Boston, September, 1865*. Boston: Wright & Potter, 1865, p. 135.
31 Based on an article in the *Worcester Sunday Telegram*, October 13, 1918.
32 Worcester, Massachusetts, Superior Court Records, November, 1860, Isaac Fiske vs. Matthew Arbuckle.
33 Worcester County, deed records, Vol. 385, p. 495.
34 *Worcester Palladium*. October 13, 1841.
35 *Reports of the First Exhibition of the Worcester County Mechanics' Association at the Nashua Halls in the City of Worcester, September, 1848*. Worcester: Henry J. Howland, 1848, p. 34.
36 *Reports of the Second Exhibition of Worcester County Mechanics' Association at the Halls on Union Street in the City of Worcester*. Worcester: Tyler Hamilton, 1849, p. 12.
37 *Reports of the Third Exhibition of Worcester County Mechanics' Association at Halls on Exchange Street in the City of Worcester*. Worcester: C. Buckingham Webb, 1851, p. 55.
38 Dodworth, p. 23.
39 Charles G. Washburn. *Industrial Worcester*. Worcester: The Davis Press, 1917, p. 299.
40 *Worcester Palladium*. June 21, 1854, Col. 1, p. 4.
41 *Reports of the Fourth Exhibition of the Worcester County Mechanics' Association Held at Mechanics Hall in the City of Worcester, September 17, 1857*. Worcester: Henry J. Howland, 1857, p. 25.
42 Examples are found in the Shrine To Music Collection, Vermillion, South Dakota, and in the Collection of Mr. Fred Benkovic, Wauwatosa, Wisconsin.
43 Phillip Bate. *The Trumpet and Trombone*. London: Ernest Benn Ltd., 1966, pp. 160, 161.
44 Florence T. Allen (compiled by). "Worcester House Directory Arranged by Streets, 1872." (Unpublished Manuscript, American Antiquarian Society, Worcester, 1954.)

Appendix

Instruments Made by J. Lathrop Allen

Key Bugle in e-flat. Allen & Co., Sturbridge, Massachusetts, ca. 1839, copper with brass trim, nine keys, Old Sturbridge Village Collections, Sturbridge, Massachusetts. (Figure 13)

Trumpet in B-flat. J. Lathrop Allen, No. 16 Court Square, Boston, ca. 1843, brass, three Vienna twin-piston valves, Collection of Dr. Thomas R. Beveridge, Rolla, Missouri. (Figure 14)

Tenor Horn with B-flat and C Slides. Allen & Co., Norwich, Connecticut, ca. 1846, copper with brass trim, three Vienna twin-piston valves for the left hand, Henry Ford Museum, Dearborn, Michigan. (Figure 15)

Cornet in a-flat, overshoulder. Made by J. Lathrop Allen, 17 Harvard Place, Boston, for H. B. Dodworth, New York, ca. 1855, German silver with five Allen valves, Henry Ford Museum, Dearborn, Michigan. (Figure 20)

Cornet in B-flat, circular. J. Lathrop Allen, No. 17 Harvard Place, Boston, 1853-56, German silver, four Allen valves, Kurt Stein Collection, Springfield, Pennsylvania.

Tenor or Alto Horn in E-flat. J. Lathrop Allen, 17 Harvard Place, Boston, 1853-56, German silver, four Allen valves, Stearns Collection, Ann Arbor, Michigan.

Baritone in B-flat. J. Lathrop Allen, 17 Harvard Place, Boston, ca. 1855, brass, three Allen valves, Fred Benkovic Collections, Wauwatosa, Wisconsin. (Figure 19)

Baritone in B-flat, overshoulder. J. Lathrop Allen, 17 Harvard Place, ca. 1855, German silver, four Allen valves, Henry Ford Museum, Dearborn, Michigan.

Valve Bugle in e-flat, overshoulder. Allen Mfg. Co., 18 Harvard Place, Boston, ca. 1859, silver or silver plate, three Allen valves, Don Essig Collection, Central Missouri State University, Warrensburg.

Valve Bugle in e-flat. Allen Mfg. Co., Harvard Place, Boston, ca. 1859, German silver, three Allen valves, Moravian Museum, Old Salem, North Carolina. (Figure 22)

Two Valve Bugles in e-flat, overshoulder. Allen & Hall makers, 334 Washington Street, Boston, ca. 1861, German silver, three Allen valves, Fred Benkovic Collection, Wauwatosa, Wisconsin.

Cornet in B-flat. Allen & Hall, 334 Washington Street, Boston, ca. 1861, brass, three Allen valves, Don Essig Collection, Central Missouri State University, Warrensburg.

Tuba in E-flat, overshoulder. Allen & Hall, 334 Washington Street, Boston, ca. 1861, brass, four Allen valves, Henry Ford Museum, Dearborn, Michigan. (Figure 23)

Cornet in e-flat and f. Made by J. Lathrop Allen, No. 111 East 18th Street, New York, ca. 1870, silver plated, four Allen valves, Loyd Davis Collection, Prairie Village, Kansas. (Figure 25)

French Horn. J. L. Allen, maker, New York, ca. 1870, brass with German silver trim, three Berlin valves, New York Metropolitan Museum of Art. (Figure 24)

Instruments Made by E. G. Wright

Ophicleide in C. E. G. Wright, Roxbury, Massachusetts, ca. 1839, brass, nine keys (mouth-pipe missing), Don Essig Collection, Central Missouri State University, Warrensburg. (Figure 28)

Key Bugle in e-flat. E. G. Wright, Boston, ca. 1845, copper with brass trim, nine keys, Dale Music Company Collection, Silver Spring, Maryland.

Key Bugle in e-flat. E. G. Wright, Boston, ca. 1845, copper with brass trim, case, ten keys, Henry Ford Museum, Dearborn, Michigan. (Figure 29)

Key Bugle in e-flat. E. G. Wright, Boston, ca. 1845, copper with brass trim, ten keys, The Music Museum, Deansboro, New York.

Key Bugle in e-flat. E. G. Wright, Boston, ca. 1845, copper with German silver trim, ten keys, Greenleaf Collection Interlochen, Michigan.

Key Bugle in e-flat. E. G. Wright, Boston, ca. 1845, copper with silver trim, ten keys, William Gribbon Collection, Greenfield, Massachusetts.

Key Bugle in e-flat. E. G. Wright for Graves & Co., Boston, ca. 1851, copper with German silver trim, ten keys, Browning Memorial Museum, Rock Island, Illinois.

Key Bugle in e-flat. E. G. Wright, 8 Bromfield Street, Boston, 1845-47, copper with German silver trim, eleven keys, Loyd Davis Collection, Prairie Village, Kansas.

Key Bugle in e-flat. E. G. Wright, Boston, ca. 1850, copper with brass trim, eleven keys, Rhode Island Historical Society, Providence, Rhode Island.

Key Bugle in e-flat. E. G. Wright, Boston, ca. 1850, silver-plated copper, eleven keys, Henry Ford Museum, Dearborn, Michigan.

Key Bugle in e-flat. E. G. Wright, Boston, ca. 1850, eleven keys, silver with gold nameplate, elaborately engraved, Williams College, Williamstown, Massachusetts.

Key Bugle in e-flat. E. G. Wright for Allen & Co., Boston, "Thomas B. Harris, Xenia, Ohio," silver, decoratively engraved, eleven keys, case, Fred Benkovic Collection, Wauwatosa, Wisconsin.

Key Bugle in e-flat. E. G. Wright, Boston, 1853, "presented to S. Wells Phillips, leader of the Greenport Brass Band, as a mark of respect by his fellow citizens of Greenport," silver, decoratively engraved, eleven keys, Greenleaf Collection, Interlochen, Michigan.

Key Bugle in e-flat. E. G. Wright, Boston, "presented to D. Chase by the inhabitants of Clinton and the Clinton Brass Band, January, 1854," silver, decoratively engraved, eleven keys, Fred Benkovic Collection, Wauwatosa, Wisconsin.

Key Bugle in e-flat. E. G. Wright, Boston, "presented to A. R. Fitch by the members of the Fitchburg Cornet Band, March, 1854," silver, decoratively engraved, eleven keys, Dale Music Company Collection, Silver Springs, Maryland.

Key Bugle in e-flat. E. G. Wright, Boston, "presented to

Rufus Pond, leader of the Milford Brass Band, 1855," silver, decoratively engraved, twelve keys, Sousa Collection, Urbana, Illinois.

Key Bugle in e-flat. E. G. Wright, Boston, ca. 1861, "presented to G. R. Choate, leader, 35th N.Y.V. Regimental Band, by its members," silver, decoratively engraved, gold nameplate, twelve keys, Loyd Davis Collection, Prairie Village, Kansas.

Key Bugle in e-flat. E. G. Wright, Boston, "J. C. Green, Providence, November 5, 1850," silver with gold-plated keys and trim, twelve keys, Rhode Island Historical Society, Providence, Rhode Island.

Key Bugle in e-flat. E. G. Wright, No. 115 Court Street, Boston, 1848-52, silver, decoratively engraved, gold nameplate, case, twelve keys, Henry Ford Museum, Dearborn, Michigan. (Figure 26)

Key Bugle in e-flat. E. G. Wright, Boston, "presented to D. C. Hall Esq. by the members of the Lowell Brass Band, April 15, 1850," gold, decoratively engraved, twelve keys, Henry Ford Museum, Dearborn, Michigan. (Figure 27)

Valve and Key Bugle in e-flat, overshoulder. E. G. Wright, Boston, ca. 1853, German silver, five keys and three rotary valves, Henry Ford Museum, Dearborn, Michigan. (Figure 34)

Trumpet in F. Wright & Baldwin, Boston, 1845, brass, three Vienna twin-piston valves, Smithsonian Institution, Washington, D. C. (Figure 32)

Valve Bugle in e-flat. E. G. Wright, Boston, ca. 1855, German silver, three rotary valves, Smithsonian Institution, Washington, D. C. (Figure 36)

Valve Bugle in e-flat. E. G. Wright, Boston, "presented by the members of the 3rd Brigade Band, 3rd Division, 9th A.C. to William Critchly, Jr., leader, July 4, 1863," German silver, three rotary valves, Smithsonian Institution, Washington, D.C. (Figure 36)

Cornet in B-flat. E. G. Wright, Boston, ca. 1865, silver, three rotary valves, Henry Ford Museum, Dearborn, Michigan.

Cornet in B-flat. E. G. Wright, Boston, ca. 1860, brass, three rotary valves, Fred Benkovic Collection, Wauwatosa, Wisconsin.

Cornet in B-flat, circular. E. G. Wright, Boston, ca. 1860, brass, three rotary valves, Henry Ford Museum, Dearborn, Michigan. (Figure 37)

Cornet in B-flat, overshoulder. E. G. Wright, Boston, ca. 1860, German silver, three rotary valves, Fred Benkovic Collection, Wauwatosa, Wisconsin.

Cornet in B-flat. Wright, Gilmore & Co., Boston, 1867, German silver, three rotary valves, John Brookfield Collection, Concord, New Hampshire.

Tenor or Alto Horn in E-flat. E. G. Wright, Boston, ca. 1865, brass, three rotary valves, Don Essig Collection, Central Missouri State University, Warrensburg. (Figure 38)

Tenor or Alto Horn in E-flat, overshoulder. E. G. Wright & Co., Boston, 1864-66, German silver, three rotary valves, Fred Benkovic Collection, Wauwatosa, Wisconsin.

Tenor or Alto Horn in E-flat. E. G. Wright, Boston, ca. 1865, German silver, three rotary valves, Henry Ford Museum, Dearborn, Michigan. (Figure 39)

Tenor or Alto Horn in E-flat. Wright, Gilmore & Co., Boston, for W. & W. Mfg. Co., 1867, silver-plated brass, three rotary valves, Henry Ford Museum, Dearborn, Michigan. (Figure 42)

Tenor Horn in B-flat. Wright, Gilmore & Co., Boston, German silver, three rotary valves, Collection of Alfred F. Wood, Westerly, Rhode Island.

Baritone Horn in B-flat. E. G. Wright & Co., Boston, 1st Div. Corp., German silver, three rotary valves, Collection of Alfred F. Wood, Westerly, Rhode Island.

Baritone Horn in B-flat. E. G. Wright, Boston, ca. 1865, silver-plated brass, three rotary valves, Henry Ford Museum, Dearborn, Michigan.

Bass in B-flat. E. G. Wright, Boston, ca. 1865, "presented to Daniel B. Davis by the non-commissioned officers and privates of Co. I, 24th Regt., Wis. Vol. Inf.," German silver, four rotary valves, Fred Benkovic Collection, Wauwatosa, Wisconsin.

Tuba in E-flat, overshoulder. E. G. Wright, Boston, ca. 1865, German silver, four rotary valves, Stearns Collection, Ann Arbor, Michigan. (Figure 40)

Tuba in E-flat. Wright, Esbach & Hartman, Boston, 1864-66, nickel-plated brass, three rotary valves, Janssen Collection, Claremont, California.

Instruments Made by Isaac Fiske

(All are signed "Isaac Fiske, Worcester, Mass.")

Valve Bugle in e-flat, upright. German silver, three twin-piston Vienna valves, half-step valve first, c. 1848, Fred Benkovic Collection, Wauwatosa, Wisconsin. (Figure 43)

Valve Bugle in e-flat, circular. German silver, three rotary valves, half-step valve first, c. 1850, Fred Benkovic Collection, Wauwatosa, Wisconsin. (Figure 44)

Valve Bugle in e-flat, overshoulder. Brass, three rotary valves, painted wood case, c. 1852, Henry Ford Museum, Dearborn, Michigan. (Figure 45)

Valve Bugle in e-flat. German silver, three rotary valves, C. 1852, Henry Ford Museum, Dearborn, Michigan. (Figure 46)

Valve Bugle in e-flat, overshoulder. German silver, three rotary valves, and one key, c. 1852, Fred Benkovic Collection, Wauwatosa, Wisconsin. (Figure 47)

Valve Bugle, overshoulder. Silver-plated brass, three rotary valves, c. 1855, Worcester Historical Society Collection, Worcester, Massachusetts.

Valve Bugle in e-flat, circular. German silver, four rotary valves, the additional valve lowers the instrument one whole-step, c. 1855, Robert E. Sheldon Collection, Washington, D.C.

Valve Bugle in e-flat. Nickel-plated brass, three rotary valves, c. 1855, Smithsonian Institution, Washington, D.C.

Cornet in f and e-flat, circular. German silver, five rotary valves, additional valves provide whole-step-and-two-and-one-half-step additions, c. 1860, Henry Ford Museum, Dearborn, Michigan. (Figure 48)

Cornet in B-flat. German silver, four rotary valves, c. 1860, Alfred F. Wood Collection, Westerly, Rhode Island.

Baritone in B-flat. German silver, three rotary valves, c. 1855, Henry Ford Muesum, Dearborn, Michigan. (Figure 50)

Baritone in B-flat, overshoulder. German silver, three rotary valves, c. 1860, Sousa Collection, University of Illinois, Urbana.

Valve Bugle in e-flat, overshoulder. German silver, three rotary valves, c. 1858, Michael Zadro Collection, New Paltz, New York.

Baritone in B-flat, overshoulder. German silver, four rotary valves, two-and-one-half-step fourth valve, c. 1860, Don Essig Collection, Central Missouri State University, Warrensburg. (Figure 58)

Cornet in e-flat. German silver, three rotary valves, Fiske arrangement, c. 1867, Greenleaf Collection, Interlochen, Michigan.

Valve Bugle in e-flat, circular. German silver, three rotary valves, c. 1870, Smithsonian Institution, Washington, D.C.

Valve Bugle in e-flat. German silver, three rotary valves—Fiske arrangement, but with spring casings below the valves, c. 1870, Alfred F. Wood Collection, Westerly, Rhode Island. (Figure 54)

Cornet in e-flat. German silver, three rotary valves—Fiske arrangement, but with spring casings below the valves, c. 1870, Robert E. Sheldon Collection, Washington, D.C.

Cornet in e-flat. Brass, three rotary valves, Fiske arrangement, c. 1870, Janssen Collection, Claremont, California.

Cornet in e-flat. German silver, three rotary valves, Fiske arrangement, c. 1870, Shrine To Music Museum, Vermillion, South Dakota.

Cornet in e-flat. German silver, three rotary valves, Fiske arrangement, c. 1875, Sousa Collection, University of Illinois, Urbana.

Cornet in e-flat. German silver, three rotary valves, Fiske arrangement, c. 1880, Henry Ford Museum, Dearborn, Michigan. (Figure 52)

Cornet in B-flat. Silver-plated brass, three rotary valves, Fiske arrangement, c. 1875, Alfred F. Wood Collection, Westerly, Rhode Island.

Cornet in B-flat (Unsigned). Brass, three rotary valves, Fiske arrangement, c. 1870, Worcester Historical Society Collection, Worcester, Massachusetts.

Cornet in B-flat. Silver-plated brass, three rotary valves, Fiske arrangement, c. 1867, Rhode Island Historical Society, Providence, Rhode Island.

Cornet in B-flat. German silver, rotary change of key valve to A, three rotary valves, Fiske arrangement, c. 1870, Martin Lessen Collection, Rochester, New York.

Cornet in B-flat. German silver, three rotary valves, Fiske arrangement c. 1875, Dr. G. Norman Eddy Collection, Cambridge, Massachusetts.

Cornet in B-flat. German silver, three rotary valves, Fiske arrangement, c. 1875, Sousa Collection, University of Illinois, Urbana.

Alto Horn in E-flat. German silver, three rotary valves, c. 1870, Don Essig Collection, Central Missouri State University, Warrensburg.

Baritone in B-flat. German silver, three rotary valves, Fiske arrangement, c. 1870, Dr. Robert M. Rosenbaum Collection, Scarsdale, New York. (Figure 53)

Cornet in B-flat. Nickel-plated brass, three Perinet piston valves, c. 1883, Henry Ford Museum, Dearborn, Michigan. (Figure 59)

Addenda to Second Printing, February, 1981

Key Bugle in e-flat. E.G. Wright for Graves & Co., Boston, c. 1855, silver, twelve keys, Michigan Public Museum, Grand Rapids, Michigan.

Key Bugle in e-flat. E.G. Wright, Boston, 1853, "Presented to Miller Cook Leader of the L. Abbington Brass Band by the citizens and members of the band," silver with gold nameplate, eleven keys, Yale University Collection of Musical Instruments, New Haven, Connecticut.

Key Bugle in e-flat. E. G. Wright, Boston, c. 1855, silver, decoratively engraved, Camden-Rockport Historical Society, Camden, Maine.